The Ferocious Summer

THE FEROCIOUS SUMMER

Palmer's penguins and
the warming of Antarctica

MEREDITH HOOPER

To dear Charlotte
I'm been so lucky to have your
friendship — and enjoy your conversation
much love
Rendell

1 July 2007

P

PROFILE BOOKS

First published in Great Britain in 2007 by
PROFILE BOOKS LTD
3A Exmouth House
Pine Street
London EC1R 0JH
www.profilebooks.com

Copyright © Meredith Hooper, 2007

1 3 5 7 9 10 8 6 4 2

Typeset in Garamond 3 by MacGuru Ltd
info@macguru.org.uk

Printed and bound in Great Britain by
Clays, Bungay, Suffolk

The moral right of the author has been asserted.

A CIP catalogue record for this book is available from the British Library.

Hardback ISBN 978 1 84668 008 3
Paperback ISBN 978 1 84668 034 2

The paper this book is printed on is certified by the © 1996 Forest Stewardship
Council A.C. (FSC). It is ancient-forest friendly. The printer holds FSC chain of custody
SGS-COC-2061

FSC
Mixed Sources
Product group from well-managed
forests and other controlled sources
Cert no. SGS-COC-2061
www.fsc.org
© 1996 Forest Stewardship Council

O God! that one might read the book of fate,
And see the revolution of the times
Make mountains level, and the continent –
Weary of solid firmness – melt itself
Into the sea! ...

King Henry IV, *Part 2, Act 3, Scene 1*

To Richard

Contents

Northern
Antarctic Peninsula,
Anvers Island inset

This map uses satellite images to indicate the mountains, glaciers and ice shelves of the northern Antarctic Peninsula. The fringing ice shelves that used to thicken the outline of the top part of the peninsula have disintegrated, making earlier maps redundant. The inset gives greater detail of Anvers Island and part of the peninsula's west coast. On the following spread there is an up-to-date map of the southern part of Anvers Island, and Palmer Station and its islands. Norsel Point is shown as an island, no longer joined to the mainland of Anvers as it was during the ferocious summer of 2001–02.

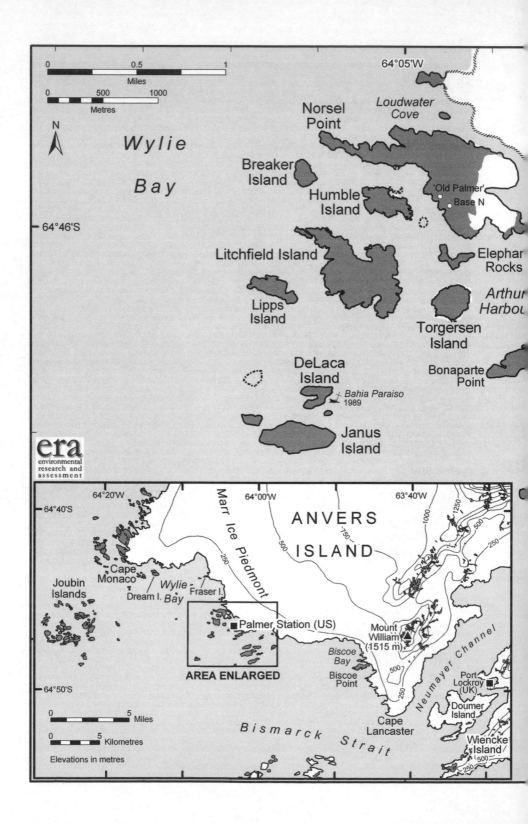

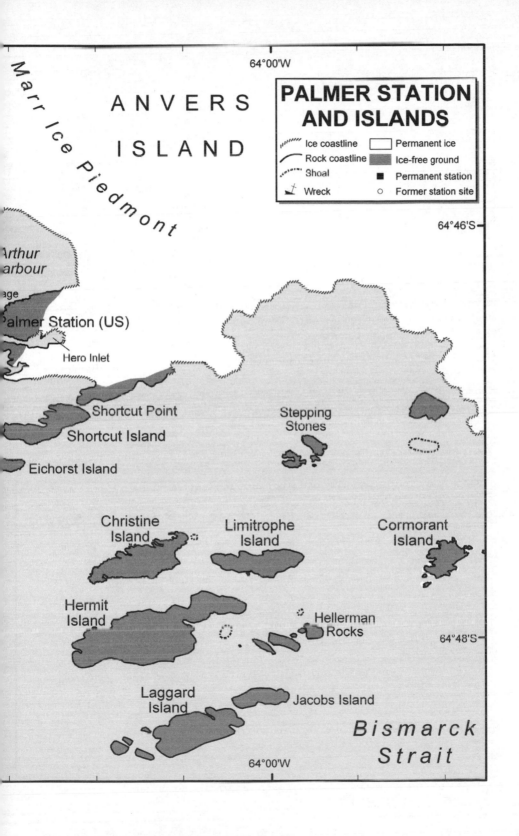

ANVERS ISLAND

Marr Ice Piedmont

PALMER STATION AND ISLANDS

Ice coastline	Permanent ice
Rock coastline	Ice-free ground
Shoal	Permanent station
Wreck	Former station site

64°00'W

64°46'S

Arthur Harbour

age

Palmer Station (US)

Hero Inlet

Shortcut Point

Shortcut Island

Eichorst Island

Stepping Stones

Christine Island

Limitrophe Island

Cormorant Island

Hermit Island

Hellerman Rocks

64°48'S

Laggard Island

Jacobs Island

Bismarck Strait

64°00'W

Introduction

News from the front line

The first explorers and scientists arriving in Antarctica carried guns to guard against potential polar bears, undiscovered tribes; at the very least, fierce somethings. But there were no polar bears in this great white weight, this unseen fridge magnet gripping the underside of the world. No indigenous people occupied any part of this coldest, driest, highest, windiest continent. The largest year-round inhabitants on land turned out to be scurriers amongst detritus, lurkers and leapers – midges, mites, springtails, tardigrades, some with antifreeze chemicals in their cells to get through the winter. No humans live permanently in Antarctica, or ever have. Anything more than half a centimetre long on Antarctica is transient.

The doing of science is the raison d'être for working on the continent and its islands, and has been ever since the Antarctic Treaty came into force in 1961. But all the labs, sleeping quarters, recreational areas, all tents, huts, stores, water-making facilities, rubbish dumps and abandoned buildings, all docks, roads, helicopter landing pads, satellite dishes and aircraft runways belonging to all the science stations and field camps in Antarctica, take up no more space than one mid-size American town.

Antarctica's human visitors arrive, as they always have, in clearly defined categories graded by function and expertise. Captains and crew. Tourists and guides. Expedition leaders and members. Adventurers and those supporting them. The four thousand or so people who live each summer in the science facilities, cut and shuffle into more intricate hierarchies: summer-only and winterers, old hands and first-timers, staff and scientists, bureaucrats having turns and briefly appearing VIPs. And, on a few stations, one, perhaps two professional humanities people competitively selected by polar humanities programmes, outsiders inserted into small communities of insiders. Eager, questioning, needing and using scarce resources. Which is the way I've come to Antarctica.

This book is about the body's fever reported from an extremity – the Banana Belt, that half-joking half-affectionate name for the Antarctic Peninsula, precipitous mountain peaks and crowding islands flicking north from the bulk of the continent. A region 2,000 kilometres long, 100 to 130 kilometres across, average 1,500 metres high. The Antarctic Peninsula is currently warming six times faster than the average for the planet, a rise in temperature more sustained than the other two known regions of abrupt climate change on Earth, central Siberia and Alaska. But on the western side of the Antarctic Peninsula the rise is ten times the mean rate of global warming.

I first came to the peninsula's west coast in 1998–99. A month with plate tectonic scientists on a US research ship, half of another month with the British Navy and two more at that scrap of US responsibility, the smallest of America's three Antarctic bases, Palmer Station. One short gravel road, a few blue-painted metal buildings, two white fuel tanks, a fist-shaped dock, five or six rubber zodiacs tethered to the shore, a summer population of nesting Adélie penguins and around forty humans, with assorted seals and other seabirds, living on a scatter of brown islands and points of rocky land separated by ocean swells, swept by fast-moving weather systems. A blip on the great white emptiness.

The chief scientist at Palmer was seabird ecologist Dr Bill Fraser, one of an inner group of US scientists who have dedicated themselves to Antarctic research. Bill's work, begun in the 1970s with modest aims, was now linking the lives of the small Adélie penguins nesting at Palmer with climate change. During benign summer days Bill explained to me how the shift in his research had happened: he'd been too busy to set down his thinking.

I wanted to come back to Palmer. I'm not a scientist. I'm an outsider to the way scientists think and work. Like most of us. But Palmer's penguins were a route I could follow into the complex business of Earth's changing climate. The Antarctic Peninsula was warming fast. Why was the warming happening? What were scientists doing to understand it? I'd been introduced to climate change five summers before, travelling along the massive eastern bulge of the Antarctic continent as a writer on an Australian research ship. Warming wasn't reckoned to be affecting the bulk of the continent. But here on the Antarctic Peninsula rocks lay naked, newly revealed, stripped of snow. Glaciers withered, ice shelves were retreating, ecologies changing. Glaciers don't have political agendas, nor do penguins. There was no debate. In this remote, austere, beautiful place, our planet was visibly heating up.

Climate change isn't a blanket thrown evenly over the surface of Earth. Its impacts are variable. They can be specific, local. Climate change can deliver sudden blows, or glancing whacks. It can insinuate – barely noticeable – then unleash a tumble of events. What changes is quantity, quality, intensity. Adjustments to the delivery systems: rain increasing, or failing, harder frosts, more snow, winds blowing from unlooked for directions, or with greater ferocity. Temperatures ratcheting upwards. Seasons shifting later, earlier. Regional spikes and lows. General accounts of climate change don't give much space to Antarctica. It's too vast, distant, unknown. Data is sparse, evidence difficult to interpret. Maps showing impacts of climate change across the world regularly leave off entirely the southern continent, this tenth of the planet's land surface. I keep a small elegant glass globe on my desk, continents standing clear of the etched oceans, the base sliced away so the globe doesn't roll: Antarctica totally missing. A constant reminder. Antarctica is central to the processes of global warming, and central to any understanding of those processes. Antarctica matters.

Back home in London I applied jointly with Bill Fraser to the US National Science Foundation Artists & Writers Program for a second grant, to research and write a book at Palmer. I negotiated with my busy husband the unexpected prospect of being seriously unobtainable. Again. I postponed commitments. Our three (grown-up) children persisted in insisting that I was abandoning them in favour of penguins. The NSF grant came through, sending me south for the summer of 2001–02.

But the chance of timing took the story I'd come to write and shoved it into the eye of climate change. The daily lives of Palmer's few thousand Adélie penguins were becoming central evidence: news from the front line. Pieces of the climate change jigsaw – that complex of interlocking pieces, bits jumbled, many missing – began clattering into place.

The Ferocious Summer is an eye-witness account of those Palmer months, a record of science happening, of ideas pitting against unfolding reality as climate change came thumping at the door. But the science of climate change is new and fast-moving. Understanding hesitates, jerks forward, stalls. Scientists regroup. Politics are integral, passions intense. The progression of ideas is a vital part of the global warming story. Scientists working on the Antarctic Peninsula needed to process data, assess what had taken place during that complex 2001–02 season, initiate follow-up projects. Conclusions took time to bed down. The peninsula continued warming. The ferocious summer required a longer chronology, and that wider context enfolds my narrative.

This book is based on my daily diaries, personal observations, and interviews both noted and recorded, in and out of Antarctica. The science has necessitated extensive reading amongst the journal literature; the history has meant archive research, interviews, and reading in the copious Antarctic and climate change sources. Workshops and conferences added to my education. Quotations throughout the book are either verbatim, indicated by quotation marks, or as reported speech, on occasion reducing or re-ordering conversations for clarity. I describe what I saw. I report what I heard, and discussed, in conversations and interviews. Selection and interpretation are by definition personal. I write about a wide and complex territory and the checking of my text by scientists has been invaluable. I have tried to minimise errors. The opinions expressed in the book are generally either attributed, or my own.

The intellectual and field time Bill Fraser gave me at Palmer is central to the narrative. Seabird ecology team members Donna Patterson, Chris Denker, Heidi Geisz and Brett Pickering provided essential conversation and guidance. My work on the Antarctic Peninsula was enabled by two grants from the US National Science Foundation Artists & Writers Program; in East Antarctica by the Australian Antarctic Division, and on HMS Endurance by the British Admiralty. Time in Cambridge as a Visiting Scholar at both Wolfson College, and Scott Polar Research Institute, gave me much needed space for research and writing.

Acknowledgement of the assistance I have received from the wider Antarctic and academic communities – individuals, and institutions – and my gratitude to those scientists who read and checked my text, is included at the end of the book, along with a highly culled selection of publications.

Three final notes: as an Australian I know in my being that summer happens at the end of the year. But northern hemisphere people can feel challenged. The summer season in Antarctica begins in October and ends with the slide into autumn the following March. It spans two calendar years, and is always described to reflect that reality, ie summer 2001–02.

I use American terminology when describing the spaces where penguins nest, because at Palmer Station on the Antarctic Peninsula I worked with American scientists. So 'colony' is a discrete collection of breeding groups with discrete boundaries; 'rookery' means the complete collection of colonies. For UK scientists the whole area is a colony, and the collections of breeding groups with discrete boundaries are nest groups.

I use metric measurements throughout the book, following the practice of scientists; but on occasion, when they seem more natural, feet and inches.

1

Ice Age

Daylight shrinks, and the sea's surface starts to freeze. Grease ice forms like an oily skim on the surface, and grey, thin, flat circles of pancake ice, edges agitated upwards. Angular chips and fragments of ice clutter together in loose conglomerations. Nets of ice crystals dampen the swells, still plastic, heaving slowly.

Brief autumn turns to bitter winter. The ice thickens and spreads, shifting, buckling, breaking apart, grinding together. Floes drift with the winds; icebergs bear down, driven by ocean currents. The ice tempts and punishes. It can open enough to let a ship through; it can pincer together with unrelenting, crushing force.

By early spring the ice surrounding the continent of Antarctica is at its greatest extent, ice-covered land merging into ice-covered sea, a doubling of glaring white. Sometimes the ice lies in flat, smooth surfaces like wet beach sand. Sometimes it separates into floes, archipelagos of ice, with still, black water between. The ice can be whipped into peaks like an endless white meringue, or heap slab on slab or jumble into contorted piles – textured, sculpted, glittering, subtle, a hundred shades of white, pierced with intense electric blue. The mind is easily fooled, and all sense of the deep, dark depths below vanishes.

Summer comes. The ice breaks apart, melts, decays, reducing in area by perhaps four-fifths. Until sun retreating, summer over, the sea's surface begins once again to freeze.

But the sea no longer freezes around Antarctica with the same grand, predictable gesture. The glittering, fantastical, white ice world is changing. Sea ice, that mix of frozen sea water and all the bits of frozen fresh water from the land – lumps of ancient glaciers, pieces of ice that once clung to the highest mountains – is in some places declining. Estimates vary for the extent and thickness of the great annual pulsing, the timing of advance and retreat, depending on the region. But along the western side of the Antarctic Peninsula there has been a 40 per cent decrease in the mean annual sea ice extent since 1979, when reliable satellite observations became available.

Ninety per cent of our planet's ice is in Antarctica. Around 70 per cent of all our fresh water, locked in as ice. The statistics numb the mind. An ocean covers the north pole. But the south pole is in the middle of a vast continent. The ice sheets lying here, over the centre of Antarctica, are so crushingly heavy that the crust is depressed, pushed down into the mantle, flattening the bottom of the globe. The ice sheets are a roll-call of annual snowfalls settling, compressing, hardening into glacier ice, moving downwards, century after century, millennium after millennium, shoving irretrievably seawards until, at the land's edges, the ice pauses in ice cliffs, decaying and crumbling or shearing off in tilting, wallowing lumps. Or it collects in floating ice shelves which spill out across the sea, shedding their seaward faces to become Antarctica's distinctive flat-topped, vertical-sided icebergs. Segments of ice unhooking from the mother lode, the size of office blocks or airport runways.

Antarctica *is* ice, seductive, harsh, infinite in its variety and beauty – all-defining. Antarctica is the ice age. It's our recent past, the way things were, but now playing out in real time. Except down here, in the south, where no one ever lived, and no one much comes.

I came to Antarctica first in late spring, edging through the pack in a research and supply ship around the continent's east coast. Metre-thick lumps of broken floe wallowed past, milky green bulk, yellow-brown undersides streaked with phytoplankton, the ocean's grass. Emperor penguins stood apart, their heads swivelling as the ice resisted. Adélie penguins ran one behind the other across the floes or dropped onto their bellies, sliding in agitation. Crabeater seals, skins scarred with the wide bite of failed leopard seal attacks, lay with new-born pups. Whales swam in patches of open water, backs lifting briefly above the still surface. Birds flew with us all day, all night. Our helicopters landed on floes like mayflies pausing on lily pads. I walked across the sea's white surface, shallow solidity above deep oblivion. Ice defined every horizon. From the air the ship dwarfed to a small red toy, shrank to insignificance, then disappeared in the white immensity. We flew 120 kilometres over the sea ice to the continent's edge, to one of Australia's three research stations. Along the horizon, wisps of snow were lifting off the high surface of the great East Antarctic Ice Sheet like hair being tugged up, heralding a blizzard. I was in thrall to the ice: forever captured.

Our ship halted in an ice plain. South, beyond seeing, the world's largest valley glacier, the Lambert, streamed 400 kilometres long, up to 80 kilometres wide, into the Amery Ice Shelf. Scientists on board were working on a range of Antarctic climate change issues: crucially, were the vast ice sheets, Earth's freshwater storage units stacked 2 to 4 kilometres high over the Antarctic continent, growing, shrinking or stable? Three red bulldozers each towing three to four sledges, with living-van, generator, fuel, supplies, radar equipment, had just set out, crawling around the drainage basin of the Lambert Glacier and its tributaries: a 100-day traverse, covering 3,000 kilometres, the final journey of a five-year programme repeat-measuring snow accumulation rates and ice movement rates using the Global Positioning System, GPS. Marker poles had been established every 2 kilometres by a bamboo cane, every 10 by a pair – each with a numbered aluminium tag, all with a beer can pushed down on top – and every 30 kilometres by a beacon tower of used 44-gallon fuel drums welded together and wedged into a hole in the ice. 'A drum may not be empty', commented a dieso mechanic laconically. The frozen contents of Blue Thunder, the latrine, needed to go somewhere. 'At −40°C your balaclava freezes onto your face. Your hair freezes stiff. Your beard is a chunk of ice. Your snot freezes. Your lips crack. There's zip all to see up there. When the wind is blowing, everything is white. The sky, the ground, the air.' 'If I ever thought of doing it again,' another dieso said, 'I wrapped myself up in a white sheet and sat in the freezer. Then I remembered.'

Preliminary results were showing that the Lambert was growing. Increasing amounts of snow were falling at the head of the glacier system. Warm air can hold more moisture than cold, and here in Antarctica – this was 1994 – warming air was delivering more snow. Increased snow because of warming seemed counter-intuitive. But increasing snow was central to the assumption that Antarctica's stores of ice were not at risk, despite climate change. The Lambert drains 8 per cent of the ice from the East Antarctic Ice Sheet high on the continent's plateau, delivering it into the massive Amery Ice Shelf, extending 300 kilometres beyond our ship. Think, one of the sea ice researchers explained, as we gazed out over the expanse of glaring white, of an ice cube floating in a glass of whisky: the level of liquid in the glass doesn't change when the ice melts – the glass doesn't suddenly overflow. In the same way, ice already floating in the sea makes no difference to sea level. Ice shelves like the Amery were afloat, and ice calving from their fronts did not add to the height of the ocean. Only ice that comes straight off the land into the sea can contribute to that. The vast burden of ice over the Antarctic continent did not, so far, appear to be contributing to any increase in sea level.

Antarctica's capacity to create, store and disperse ice is critical to the way our planet functions. Any changes in the volume of Antarctica's land-stored ice, any changes in the annual timing and extent of its sea ice, have an impact on the ocean and the atmosphere: in the sea – on temperature, density and composition, on the currents rolling through, on living things within; in the air – on the amount of moisture it carries, on temperature, on the array of shifting winds, on their direction and speed. On the climate.

Climate is what you expect. Weather is what you get. Climate is the average weather. Climate change is getting weather we are not used to. Climate change is changes to the average, to the chances of something happening: changes in probability, increases in the frequencies of what the weather does. Climate change is when climate goes beyond what is considered natural variability. What was predictable becoming unpredictable. Uncertainty beginning to dominate.

Climate change requires some sense of what was, before what now is or may be. Are there changes, and are they measurable? Scientists need information extracted from the physical world – from visible, or hidden, evidence. They need databases of observations: regular, repetitive, reliable, collected over sufficient time periods. Climate change is about measuring. It's about the passionate determination of individuals to commit to tracking, and recording.

The quantities of greenhouse gases in Earth's atmosphere were measurably increasing. There was more methane, more nitrous oxide, Ever-increasing amounts of carbon dioxide, CO_2, were being released into the atmosphere as we burned ever-increasing amounts of fuel made from the fossilised remnants of plants that had once scavenged the carbon from the atmosphere and stored it. The plants had grown and died over hundreds of millions of years. Now we were burning their remains with such zealous persistence that we had created a significant and growing human-driven system pushing carbon dioxide emissions into the atmosphere, alongside the natural systems that had always released carbon dioxide. As the greenhouse gases increased, they trapped heat from the sun, further warming Earth's surface. Rising temperatures, the scientists on board pointed out, had huge implications to our planet. Weather would become less predictable, more violent, with increased frequency of extreme events: heatwaves and drought, heavy rainstorms, cyclones.

The rate was ratcheting up every year. The annual CO_2 emissions of any one of hundreds of coal-burning power plants now equalled the entire global output of carbon dioxide in 1800. Carbon dioxide has a long and complex lifetime. We were committing to the future's climate change, making irrevocable decisions for our children's children.

Near one of the Australian bases I'd seen a ponderous black boulder balanced on a stalk of ice, like a cumbrous seed pod on a stubby white stem. The warmth absorbed by the dark surface of the boulder was sufficient to melt the ice around it, leaving just the stalk. White is an excellent reflector of sunlight: we feel cooler in white clothes. Snow-covered ice is the best reflector on Earth, bouncing the heat reaching the planet's surface back out. The sun's heat isn't absorbed. When the sea around Antarctica freezes, most of the sun's energy is reflected back. On the scale known as a surface's albedo, spring sea ice reflects 80 to 90 per cent of the light from the sun. But the albedo of water is very low: the open ocean – in effect, black – absorbs the energy of the sun, reflecting only about 7 per cent. If a part of the ocean usually covered by ice does not receive its annual coating of white, it warms. More heat goes into the system. Less ice can form; more of the sea's surface stays unfrozen. The impact feeds on itself. If it gathers momentum, it's a positive feedback. A small start can rapidly become very big. White retreats, black grows. The best reflector becomes one of the worst. Any changes to Earth's albedo affect the amount of energy absorbed by the planet.

United States research ships carry soda, juices of various colours, but no alcohol. Australian research ships have the odd example to hand. Consider, someone said, the way chilled champagne fizzes as CO_2 escapes. Warm champagne is flat. Cold liquids can hold more CO_2 than warm liquids. Oceans are great repositories of CO_2. The Southern Ocean, almost 10 per cent of the planet's total seas, takes up around 20 per cent of the global ocean CO_2. But a warming ocean stores less CO_2 than a cold ocean.

I walked over the frozen land in one of Antarctica's rare ice-free areas, frozen lakes caught in its ancient folds, brown cliffs banded in horizontal stripes of black. This ice-free, wind-carved, naked landscape had been above the sea for only about six thousand years, and it was still rising. In Antarctica's dry cold, precipitation rates are low, and snow falling on these dark surfaces evaporated: it didn't accumulate. Three and a half million years ago dolphins had swum in warm seas where I was now walking. Their fossilised bones had recently been discovered. But the rocks themselves – they were two and a half billion years old.

It was a reminder of things we all knew. Earth's climates have always varied. Earth's surfaces have always shifted. Oceans widen, close up; continents slowly scatter; islands move; mountain ranges push high, as the massive plates they ride on grate past each other, pull apart, collide, sink. Volcanoes, earthquakes, tsunamis are acknowledged indicators and markers of the underlying tectonic pressures and strains. Predictable in their occurrence, unpredictable, violent, destructive, at the moment of occurring.

The scale of Earth's see-sawing past is vast. Palaeontologists list the 'great global dyings' when climate lurched irretrievably, estimating percentages with cool neutrality: as many as 90 to 95 per cent of all species in the fossil record vanishing during massive upheavals, biodiversity plummeting. After large-scale extinctions or extreme habitat change there is confusion, disorder in the environment and evolutionary jumps. When the climate changes, species – including humans – move, modify or perish. Adjust or fail.

Evidence of past climates is part of the litany of our living: the limestone we build with, the soils we grow our crops in, the once prolific forests we dig up and reuse as coal. The planet's colourful past is like a series of museum dioramas: woolly mammoths trumpeting icy breaths where tourists now wander with bottled water in London's Trafalgar Square; iconic dinosaurs roaming through swamps where sand now shifts over deserts. The signals of past ice ages are visible everywhere: drowned coastlines, gouged valleys, vast lakes, random boulders, caches of river gravels daily quarried to boost our concrete. In my north London garden, pebbles are scattered through the earth, yellow, red, knobbly-brown, scooped up by a glacier, then dropped here by its dirty, declining skirt. I grow plants in a layer of glacier-transported soil. Just down the hill, topsoil scraped away and with no replacement deposit, there's only unforgiving clay.

Now the climate change plot is rolling again. The curtain has already risen on the latest version of the story. This time there's new scenery, a different cast list. But the evidence of change, driven mainly by increasing temperature, is unambiguous. Ice shelves collapse, coasts disintegrate. Arctic sea ice declines at an ever-faster rate. Corals bleach or drown. Permafrost is melting. Glaciers rapidly shrink and retreat. Vital water resources are threatened. Climate zones are moving: plants are failing to manage in one place, establishing further north or south; animals are no longer foraging where they have always foraged because their prey are no longer there. Seasons lengthen or contract. Oceans warm.

Climate change is marvellous and merciless: the greatest, most important, most exciting, most serious challenge we humans face. It reaches into every part of what we don't know. It underlies much that we need to understand. It scoops more and more scientific disciplines into ever more ambitious and important attempts to comprehend global systems. It demands that we think connectedly. Trying to understand means researching how atmosphere, oceans, ice, land and all living things interrelate. It means unravelling feedback mechanisms in all their complexity. It means including in the ledger our planet's place as one component of the solar system, and the effects of known cycles on its climate – variations in Earth's orbit, the tilts and wobbles of Earth's axis. It means including the amount of heat generated by the sun, the extent to which Earth absorbs or reflects the sun's rays. It means including the slow drifting of continents and the impact of sudden random events, erupting volcanoes, material hurtling in from space – all the shocks of arbitrary catastrophes.

And it means understanding ourselves, that ongoing complexity, and how we relate to our planet. At the end of the most recent ice age – temperatures fluctuating, the swollen ocean, pumped with melting glacier ice, challenging the outflow of rivers, stealing valleys, shrinking hunting territories – there were perhaps 5 or 6 million humans. At the start of the nineteenth century the planet held a billion humans, in 1961 3 billion, in 1986 5 billion. By 2001 there were over 6 billion humans, and rising. Changes in the climate have delivered sudden, destructive shifts in their familiar world to humans in the past. People tried to manage the consequences. But they could not anticipate the cause. This time the climate plot has key new characteristics: our sheer numbers, putting extraordinary pressure on the planet's systems, and our role as part of the cause. We are knowingly contributing to the increasing loading of greenhouse gases in the atmosphere. And we know it matters. George Denton, US glaciologist, hunter of the ice ages, when I talked to him in New Zealand in 2006: 'Ice ages and the amount of carbon dioxide in the atmosphere track each other. They go hand in hand. Carbon dioxide does not drive the ice ages. We don't, so far, know what does. But currently, CO_2 is higher than it has ever been in the last 650,000 years we have so far been able to measure. It is going up too quickly. The speed is unprecedented. Global warming is real.'

Proof that atmospheric carbon dioxide was on the rise came from regular measurements begun at the south pole – Earth's least contaminated place – in 1957, during the International Geophysical Year. The rising quantities of CO_2 are memorably enshrined in an iconic graph showing the saw-toothed upward climb of the gas, measurements begun the same year in the air over the ocean-isolated mountain Mauna Loa in Hawaii. Evidence that the level of CO_2 in the atmosphere and average temperature track each other came much later, retrieved from ice cores prised out of Antarctica's massive ice sheets, time capsules of past climates, archives of our planet's atmospheres. Patient analysis of bubbles of air trapped by snow falling year after year onto the ice sheets, and the isotopic composition of the ice itself, varying with the temperature of the snow at its formation site, revealed a close link between average temperature and greenhouse gases. The strong correlation has held as researchers examine cores pulled up from ever more ancient ice. As greenhouse gas levels rose, average temperatures rose, as one fell so did the other, right back to the oldest Antarctic ice cores so far studied, from Dome C in East Antarctica, figures published in mid-2007: 800,000 years.

Much of climate change science is very recent. Scientists are under heavy pressure to try and predict what may occur, to extract information about what has already occurred. The speed of current warming forces explanations ahead of the data scientists would prefer to have, actions ahead of sufficient understanding. Predictions wobble on the cusp of inherently unpredictable future human actions, on the unknowable rate of future carbon dioxide levels. Natural variability is hard to assess. Data from new technology don't necessarily mesh with previous methodology. Scientists engage, struggle, argue, compete, make progress, regroup, recharge. They state conclusions in hedged-about, cautious phrases. Steve Haranzogo, British Antarctic Survey meteorologist: 'We have a narrow lens. We don't have a complete understanding of a complete world system. The most complex models have only been done for fifty years. We have limited measurements. We can only statistically infer from what we believe to be a valid result.'

We used to believe that climate change happened slowly, relatively smoothly, impacts stretching out, with time to accommodate and adapt. Certainly not within a lifetime. Now we know that climate change can happen with sudden slips and shifts. It can jerk unpredictably, accelerate abruptly, taking ecosystems and humans by surprise. Evidence for staggeringly fast temperature drops and rises as the northern hemisphere emerged from the last ice age have been found in a multitude of ice samples from the Greenland ice sheet. How and why these rapid, abrupt changes happened isn't really understood. Predicting when they may happen isn't yet possible. Scientists use simple images to illustrate the message of abrupt change. Push at a boulder on a hillside and it rocks, settles, rocks, settles – then careers off, bumping, smashing, uncontrolled. Press a light switch – gentle pressure, nothing, more of the same, nothing, then the light goes on, an instant shift from one state to another.

What matters, now, is trying to understand the speed of change. Research shows that abrupt climate change can occur when gradual causes put Earth systems across a threshold, switching the climate to a new state. Earth's climate system is robust. It has built-in delays. It tolerates. But there are critical points when change becomes irreversible – gateways or tipping points. An accumulation of feedback mechanisms, a series of small nudges, can amplify to a big change. The analogy with our own bodies is clear. Push a part too far, and something gives.

Time in climate change stretches or tightens: it can be counted in millions of years, or thousands, or hundreds, or decades. Or right now.

2

Punta to Palmer

30 December 2001 to 6 January 2002

Punta Arenas is the starting block and finishing line for every trip to the Antarctic Peninsula with the United States Antarctic Program. It's the knot in the long lasso. Down you go, and back you come, to the same hotels, the same places to eat. This is only my second time with the Americans, and I stay where I stayed before. The Isla Rei Jorge is smallish and comforting, and has an upholstered sunroom where Antarctic travellers pause and fidget if, as frequently happens, they are suffering delays to their plans.

The notice-board in the brown polished hall offers a festive dinner with exotic animals:

Beaver in two ways

Wild Goose

Hare

Rabbit

Rhea

Yesterday I was finishing Christmas in London. Now, 54°N to 54°S, winter to summer, the temperature is the same. Outside the windows a pack of dogs trot past, stop at the traffic lights, cross briskly together and turn right. Punta Arenas is laid out in the same grid system as a thousand former frontier towns. But here on the Straits of Magellan icy winds gust dust around street corners, and the pavements are cracked, concrete blocks heaved up by winter frosts. A scientist tells me about his girlfriend, who flew down from New York to say goodbye, bringing a big bag of treats for Punta's feral dogs – plastic bones, biscuits in dog-bone shapes, biscuits with pictures on them. The dogs ignored the plastic bones. They didn't even want the biscuits. Punta's dogs scavenge along the beaches on the edge of town, eating seafood and crabs. Local people and dogs co-exist.

Downstairs in the hotel's small dining-room two bulky Russians make breakfast sandwiches from pre-sliced white, inserting pre-sliced squares of cheese and ham stuck down with jam. The elegant bread rolls on the side table are pretend, the same as last time; but the wedges of fruit tart on flowery plates are real, and the tiny marron-filled meringues. Waiters gravely offer instant coffee. Outside, a rose is propped against the wall. Sparrows chirp, the laburnums have finished flowering in the town's central square. These are the wary, long-lit days of a short, high-latitude summer.

The R/V *Laurence M. Gould* is already in port. An orange-painted workhorse, the *Gould* makes a repetitive pattern of research and supply trips along the Antarctic Peninsula's west coast under charter to the US National Science Foundation, Office of Polar Programs. The annual January Long Term Ecological Research cruise (LTER) is about to begin, and expeditioners roam the Punta streets. Experienced Principal Investigators check their equipment in the warehouses of the local agent, Agunsa. The *Gould* will offload people and cargo at Palmer, then head south, tracking a marine biology grid, sampling the ocean to protocols begun in 1991.

At the clothing store smiley Alejandro hands me my allocation of extreme-weather gear. Trying it on is like putting on armour. I remember the bulk from the last time, the clumsiness of layers, long johns, thermal trousers, windproof trousers, feet weighed down by fleece-lined, ridge-soled, high-laced Sorrels. But I love my Sorrels: they take me where I would never dare to tread. Three years ago, when I first went to Palmer, the long johns were for men only. This time there's a version for women, so no more redundant seams and openings. But there are the same men's fit-everyone plaid shirts, and the same fly buzzing around the changing-room. At least it seems the same fly. All thirty-two allocated items must be signed for, then squeezed into a duffel-bag with my name and project number and padlocked, ready for delivery on board.

In the town square men sit impassively under the scraggy grey-green of the laburnums. Dogs watch from the thin grass. I reach up furtively and touch the polished bronze foot of one of the Indians at the base of the statue of Ferdinand Magellan, a talisman to ensure a safe return from Antarctica. Higher up the statue, the burnished breasts and bright polished nipples of a mermaid with a double tail indicate better luck. But that involves climbing.

I soak up the pleasure of a last bath and gaze for the last time at trees, bushes and real flowers, at strangers and babies. Then I walk to the *Gould*, moored at the end of the long, narrow quay. Past the small, grey, Chilean naval vessels purchased from the British Navy, past the high-prowed fishing boats, and a rusty vessel with a winch and platform at the stern capable of hauling up very large marine life. The wind blows fresh across the bay. Overnight an enormous private-balconied tour ship has arrived opposite our stubby *Gould*. The two deck cranes loading our last supplies look like Toytown. Cleaners hang in gantries down the liner's white sides, polishing the views. Passengers coming ashore claim me with friendly greetings; they think I'm someone they haven't yet met. But I'm for the other.

Jay sits inside the little wooden pillbox parked on the quay, arms colonised by tattoos, sing-song Cajun voice, knowing eyes in his nine-lives face. But his pepper-and-salt, shoulder-length hair is short because this time Jay is First Officer, and Rob, with the growly voice and cascading dark curls, a passion for Harley-Davidsons and hard rock, is now Captain Robert. To Captain Robert, I'm Miss Meredith. But to Jay – 'Well, hello girl,' he says. And I've arrived. A net of stores swings overhead. I wait till it clears, then climb our steep gangplank and step onto the *Gould*'s familiar metal decks, negotiate piles of scientific equipment and bags of kit in the Baltic Room, pull open the heavy watertight door at the bottom of the stairs, tackle their steepness, think about the winterer from Palmer who tumbled down them four months ago to his death, pull open the heavy watertight door at the top, find my cabin. And bump into Bill Fraser on the way to the galley, baseball cap moulded to his head, faded blue jeans, long-sleeved blue shirt, quizzical smile. As before, except this time there's no moustache. Bill speaks Spanish with the kind of accent that makes the officers of visiting Argentinian naval vessels melt. His sandy hair and pale blue eyes, his fair, thin-freckled skin and lean build reveal his Scottish ancestry. But Bill was born and brought up in Buenos Aires, and to Argentinians he is an Argentinian. Now he is an American, from Montana. But to me he is that concept of the mind, that impossible entity, a citizen of Antarctica. A doer and a thinker, independent, respected, he has come south most years, since arriving at Palmer in January 1975 as a graduate student, studying seabird populations.

We sail on 2 January. At lunch I sit next to the Chilean piloting the *Gould* through the Straits of Magellan. Swells slosh evenly along, bisecting the portholes, and he tells me about Antarctica's last volcanic eruption in December 1967, on Deception Island. Ten miles away their naval vessel rocked in the sea, and no one had a clue what was happening. He flew a small helicopter to rescue people through thick black smoke, and stones falling from the sky. The first time I walked over Deception's lava hills a lanky technician slipped a stone out of his pocket and dropped it on the purple-brown surface striped and streaked in snow. 'That'll fool a geologist one day.' And his thin beard worked in a gentle, satisfied smile at this small iconoclastic act. 'I always do it. Pick up a stone some place in the world. Drop it some place else.' Deep in Deception's caldera the volcano glows, a real pulsing eye unseen on the ocean bed.

The *Gould*'s galley has few formalities. We help ourselves to drinks and slices of pie at the island fixture in front of the serving hatch, sharp corners requiring negotiating in heavy seas. Seats nearest the serving hatch are reserved for the Captain and officers. The predominantly Filipino crew prefer the back of the galley. Otherwise we sit where we choose at the long formica tables, bottles of sauces and concoctions jammed in permanent boxes down the centre of each, revolving chairs fixed to the floor. Bottles of sauces are cultural hazards. I'm wary of them. Newly elected as food representative at my graduate college in Oxford, I queried the need for the range of sauces on the table at breakfast. A solid young man who went on to be a professor of politics explained, forcefully, to this ignorant Australian that they were his right. I don't even touch these American signifiers.

Seven hours after leaving Punta a launch approaches from the distant coast to collect our pilot. The light is flat and grey, the wind whippy. The pilot steps across the water down on to the lurching roof of the cabin, a small, upright man in a dark overcoat with a backpack. Then he looks up and waves, and waves. Australia's research and supply ship *Aurora Australis* leaves Hobart for Antarctica with a brass band playing on the quay and coloured paper streamers snapping in mid-air as the ship pulls away. Celebratory, symbolic, long-established rituals of departure. The *Gould*'s departures from Punta are workaday, unacknowledged. I'm grateful for the pilot's courtesy. Ships need farewells.

We are on our own. There are no passports where we are going. No police, no government. No shopping, except a small station store opening briefly for shampoo, booze, postcards, T-shirts and souvenirs. There's no television or radio. No cars, no pets, no theatre. No suits. No wasps. In fading light, and rising seas, we head south.

Every five or so days storms whip anticlockwise around Antarctica, funnelling through the 900-kilometre-wide gap where South America and the peninsula veer away from each other as if punched apart. Yachties can wait to slip across the unloved Drake Passage, the fastest-flowing current in the southern hemisphere. Tourist ships travel at well-stabilised speed. But we are slower, and we go when we are scheduled. The Drake is to be endured. Three years ago I spent a month in the *Gould* as it trundled around the Drake and the Bransfield Strait between islands and peninsula, while geophysicists measured tectonic plate activity with GPS receivers. We held station, reverse thrusters grinding and throbbing as yellow ocean-bottom seismographs were deployed, then cross-hatched against the run of the swell, backwards and forwards, recording exact position. I worked in a narrow portholeless laboratory, bracing on a high stool, my laptop strapped to the bench. I know this ship.

Transit trips aren't comfortable. The *Gould* is built to push through ice, carry loads and do science. With a high, heavy bow, intended length truncated by costs, and a permanent list, it combines pitch and roll with a final sickening twist to the stern. Beyond the heavy sea doors the wind-scoured decks heave and tilt. In the galley people pick listlessly at dry crackers and drink soda out of paper cups – this season's panacea for seasickness.

The night before arriving at Palmer, Bill gives me a briefing. We sit squeezed between bags of kit in the small cabin he shares with his partner and co-worker, Donna Patterson. Finding a place to talk on the *Gould* isn't easy. Videos dominate the lounge; the bridge is cramped. People are ready to start work and haven't. Most are just passing the time, or feeling sick. But Bill can't stop. He has been in the field almost continually for over a year – last summer at Palmer, followed by a slug of winter cruises in the biologically rich Marguerite Bay area south of Palmer.

The news is shocking. The season, Bill says flatly, has gone to hell. Palmer's Adélies are in crisis, barely holding on. The weather has been relentless, dire. The seabird work is under real pressure. 'We are arriving to a catastrophe, walking into a bitter scenario produced by climate change. The Adélies don't have the capacity to survive the drastic changes that are occurring. There's no doubt.'

The real penguin losses in Antarctica are happening on the Antarctic Peninsula, where the greatest warming is occurring. Bill describes landing on the low, ice-enclosed Dion Islands during last winter's cruise in Marguerite Bay. In 1948 21-year-old Bernard Stonehouse, surveying for the British with a husky team over dodgy sea ice, discovered an emperor penguin colony on the Dions, the furthest north these polar penguins breed. Bill arrived at dawn one August morning. Washed pink sky, pearly grey ice, soft-focus light. He found just nine lonely pairs. Since Bernard made the first studies of an estimated five hundred birds, the colony has been little visited. Outside influences can't be discounted. There can be no protective fences around vulnerable bird populations to exclude helicopters or passing yachts. But for Bill, the sight of those remnant pairs of emperors at a critical period of their brooding phase was deeply symbolic. 'It was the saddest sight. They won't survive. They were the only known emperor penguin colony on the Antarctic Peninsula.'

Each summer, beginning in October, the seabird team at Palmer record their data and observations. The focus is entirely on the birds of Bill's precise patch, the Adélie penguins nesting on five inner islands and a scatter of outer islands. Year after year Bill has worked to find out what affects the survival rate of the Palmer Adélies. Factors have been scrutinised over many seasons. Identified impacts have been measured, hypothesised impacts that seemed not to be relevant have been knocked away. Bill: 'Each season here at Palmer we do what we always do. Now ecology deals us a wild card: unprecedented snow. My penguins are in difficulties.' But, ever the scientist: 'In effect, a natural experiment is occurring. Hypotheses that have been in development for a long time are being tested. There will be measurable results by the end of February.'

This season will be different from anything ever recorded. I've got a lot to catch up on.

Next morning, with no warning, we slip in sideways to Antarctica.

Palmer should be approached the stylish way – the way mariners came probing in small ships, or sealers, those unofficial explorers taking quick profits from disposable owners. Or naval commanders and gentlemen explorers, seeking geographic and scientific discoveries. An arc of tough, abrupt islands arch west of the Antarctic Peninsula, parallel to the coast: Elephant and Clarence, King George, Greenwich, Livingston, Snow. Land can pass unnoticed here, hidden in mist, or reveal tantalising, low-level slices, unseen uplands and peaks truncated by cloud. But head east, past the dramatic sunken caldera of Deception Island, to the Gerlache Strait, and performance is guaranteed. Here the two big islands of Brabant and Anvers nudge close to the mainland. Mountains rise abruptly out of the calm, blue-black water, ice-clenched peaks streaming ice-clouds like the banners of samurai warriors. Ice cliffs barricade the shores. Glaciers tumble in suspended motion. Lumps of sea ice, layers stained with phytoplankton, heave and sigh. Adélie penguins ride the slopes of small, tilting icebergs. Crabeater seals rest on stationary floes in secluded inlets. A narrower channel, the Neumayer, edges between large Anvers and small Wiencke Island, named after a young sailor, Carl Wiencke, who drowned here in 1898. Refrigerator-size pieces of clear glacier ice ride low in the water. There's a kind of massive, deep silence, a sense of waiting. An enclosed place where mountains surround the ship. A final corner jostling with glacier fragments and brash ice, recent accumulation from rapidly disintegrating glaciers, enough ice soup to enmesh on occasion the ice-strengthened *Gould*. Once through, the way opens along the southern coast of Anvers Island, west, out towards the ocean. Half-way, the aerial arrays and low buildings of Palmer Station cluster on a rocky promontory. Humdrum humanity improbably sprouting in the emptiness, a crumb on the vast white tablecloth.

But we've come the boring route, straight across from the tip of South America to Anvers, saving six hours. No first iceberg floating in mid-ocean, detached, improbable. No jiggling, tinkling scum of left-over ice pieces, scattered white over blue. No exhilarating sense of power shifting as ice restrains and seas flatten. Just the short, high, queasy swells of the Drake. Since leaving the low coastlines and distant mountains of Tierra del Fuego five mornings ago we've seen no land. Now, suddenly, the *Gould* is among small, nondescript islands compressed against an indistinct mainland, and we have almost arrived.

The grey sky delivers scraps of snow, blowing horizontally. The sea is grey and lumpy. Crowded on the *Gould*'s small bridge, I can't get my bearings. I've spent weeks among these off-shore islands at Palmer, yet nothing looks familiar. But I can see one immediate, fundamental difference. This is high summer. The islands should be rocks streaked with snow. But they are snow streaked with rocks. Adélie penguins stand forlornly, scattered like freckles on skin, a smear of penguins across the landscape.

Bill stands on the bridge, tense. Absorbing the landscape he knows intimately. Noting out loud. 'There's no penguins in the water. You'd normally expect them.'

The islands wind by – Hermit, Shortcut. Litchfield – out of bounds to everyone except Bill and his team. 'The entire rookery is under snow.' Torgersen, with huge snow banks. 'See the penguins in the snow? There's hundreds. They've lost their nests. They're just hanging out. Those birds are not feeding.'

How many penguins are missing?

Bill scans a colony he knows well. 'Easily 50 per cent off. It's a disaster. All these snow banks shouldn't be there. And now there's fresh snow accumulating on the old.'

He checks the ridge tops. One skua. Normally there are a lot. But in the harbour a leopard seal relaxes on a small, sofa-shaped floe. One of Bill's field team has reported seeing a leopard eating eleven or twelve penguins a day. The floe dips and rises gently. The leopard flexes its tail, stretches each flipper and opens its wrap-around mouth wide.

A lone penguin stands on the Palmer pier, a big stolid gentoo, white patch above each eye, red beak matching the red of its felty feet. Not one of the local Adélies. Three years ago at Palmer I used to joke that gentoos were estate agents checking out potential property. Now here is a gentoo, symbolically waiting.

People appear from the BioLab building, shrugging on float coats, the regulation buoyancy jackets, pulling the safety-first beaver tails up between their legs, fanning along the rocks to handle the *Gould*'s heavy mooring ropes. We stare at them. They stare at us. Chance pushing us all together. It's Sunday, the Station day off. And it's not even nine o'clock in the morning.

Bill: 'They'll be saying, "here comes those effing red coats".' He doesn't wear one. He looks at me. 'Don't wear one.' But of course I have the standard coat issued by the National Science Foundation. And it's red.

Getting ashore in Antarctica is a negotiation. Ninety-five per cent of the coast is ice-bound, the rest rock-bound, with occasional grey beaches swept by surf. Penguins torpedo in, underwater fliers, travelling the clear water corridors. A quick swerve to the surface, snatching a sighting of distance left, height to achieve. Then, at land's edge, a leap upward to come down, on rock, ice or snow. Or fail and tumble back to try again, and again. We humans, clumsy in our clothes, climb or clamber from helicopters, aeroplanes or small boats, or occasionally, as at Palmer, the privilege of a gangplank.

I arrived on the Antarctic continent, the first time, arms full of bananas, tomatoes and lettuces – freshies for Australian expedition-ers cut off from the rest of the world for eight winter months. We flew in two small helicopters over 120 kilometres of sea ice, crossed by lines of incoming Adélie penguins heading for vast nesting sites on the continent's edge – trails like zippers where they'd tobogganed on their bellies, digging their flippers in either side.

I've walked onto the continent across 3 kilometres of fast ice, trucks driving past along a newly scraped two-lane ice highway, shifting 620 tonnes of cargo from ship to shore. Twenty thousand Adélies nested on a nearby island, and all day, and late into every evening, lines of penguins travelled to and from the distant sea. And I've walked onto Antarctica over a suspension bridge built by Chilean soldiers out of strips of wood from packing cases, each narrow strip a foot-dropping gap from the next, a single wire cable, hand height, either side. I was the kind of child who was scared of water glinting between chinks in a jetty a fin-gernail apart. The bridge linked the Chilean stations on its small island to the ice of the peninsula. Below, ice chunks clunked in the surging tide. We'd sat down to lunch in the officers' mess, served by waiters in black tie. Now our hosts wanted to show us the aircraft landing strip, a skiddoo ride a thousand feet up the ice piedmont on the mainland itself. Crossing the bridge was one thing, but it had to be re-crossed. And it swayed, so only one person could do it at a time.

Antarctica is a thousand serious things. A thousand significant things. A thousand beautiful things. But it is also the place where I found out I could do more than I ever imagined I could. Which isn't why I went, but was an extra. I'm not an adventurer, a high-achieving camper and tramper. I've never even skied. Our holiday money went

on flights between London and Australia, showing my English husband
and our half-and-half children my side of the world. But the first time
I went south, to Antarctica, I discovered unimagined capacities to do
things I never thought I could – and people willing to help me. That's
heady stuff.

Continental Antarctica turned out to have an unexpected familiarity
– long distances, dry, clear air, deserts – similarities with the part of
Australia where I grew up, the inner curve that once fitted against the
bulge of Antarctica before Australia finally broke away from the super-
continent of Gondwana 30 to 35 million years ago. Maritime Antarctica
– the peninsula and its islands with their mountain peaks thrusting up
out of the sea, their broken complex coasts and moist grey atmosphere
– has always seemed to me curiously northern-hemisphere.

Last time I combined living and working at Palmer with weeks
exploring the Antarctic Peninsula and the islands of the South Shetlands by research ship and helicopter, privileged encounters with coastlines and ice. This time I would be island-held. The nearest point on the
mainland was only 30 kilometres to the south-east. But it was another
country. Blanked out by cloud or mist, or suddenly revealed, sharply
defined mountains, transposed closer in the brilliant light, but scenery
in a parallel world we were not participants in.

We disembark onto an artificial fist of grey, cindery stones packed
inside corrugated metal walls. Palmer's pier. We stagger a little, because
even five days in the Drake means sea legs, and look back. The orange
ship squealing against the black rubber pontoons already seems an over-
large interloper. Immediately allegiance is transferred. The *Gould* is
now the outsider. We are on the inside.

Every time the *Gould* docks, new people arrive, others leave. Those
departing move straight onto the ship, into our berths, and we take their
beds. Patterns develop quickly among small, enclosed groups. Conven-
tions sprout and flourish. Who sits at which table, who likes a particular chair in the lounge, whether this season's cooks will allow others
into their spaces. The insiders are already ensconced, but departures
alter the balance. Now we need to fit ourselves in.

Palmer is Bill and Donna's domain. Their bedroom – with a quiet view of the sea – waits in Bio. A sheet of plywood stands ready to be manoeuvred upstairs around awkward corners to transform two bunks with a brief bit of floor space into a double bed with even less floor space, but at least an ex-upper bunk shelf. The only (known) double bed on Station.

Since I was last here a layer of Palmer's quirkiness has been removed. Three years ago I slept in a four-berth room opposite the exit door of the bar/lounge/video room in GWR (short for Garage Workshop Recreation): Palmer's much-loved buildings still kept their original names. We had two cooks on Station, both named Dawn. Lead summer cook Dawn had the late rising timetable. Early Dawn occupied the bunk above me, and every morning at 4.30 she clambered down to cross to the kitchen and start bread-making. The three small windows of our room had no coverings, and the hard bright light of summer poured in. We safety-pinned grey army blankets around our bunks to create an illusion of dark. I'd come to research and write a book, illustrated by American natural history artist Lucia deLeiris. Lucia constructed a blanket box around her bunk, like a seventeenth-century Dutch cupboard bed. On the mattress above, a small friendly krill researcher curled her naked self inside a down comforter and slept through the summer light.

But through that triptych of bare windows I could absorb quintessential Antarctica. A brief foreground of brown boulders fronted the sea, surface satin-smooth or ruffled blue, strewn with constantly renewing ice fragments. Across the bay, the long line of the ice cliffs. Palmer's glory. Seductively permanent, heart-stoppingly transient, each detail poised on an inevitable collapse. Vertical slices of historic snowfalls, crevassed, sagging, buttressed, leaning pillars and improbable towers, unreachable caves and teetering overhangs like timeframe-trapped curls of solid smoke. Electric blue, brilliant white. Here, at the interface with the sea, the ice was visibly active, disintegrating in small slips and isolated falls, and occasional stupendous collapses, profiles transformed as ice slides thundered into the sea and a cloud of ice crystals rose, and a new cast of ice fragments slow-motion-rolled in the impact wave.

Rising behind, powerfully omnipresent, the ice sheet clamped over the island we were clinging to the edge of, sliding inexorably towards the waiting ocean. An uncompromising hard line of dense white intersecting with the sky, no negotiation of rock or growing things to clothe the boundary. A frozen wave of water fresher than an ad man's dreams, older than reckoning, its silent white bulk a thousand feet high here, where we saw it.

Now GWR has been renovated, with neat two-person bedrooms and smart communal bathrooms, a medical suite, gym and an upstairs deck, used by smokers. This time I'm in Bio, sharing with the other foreigner: high-energy, straight-talking Steffi from Hamburg, with short blonded hair, finishing the field work for her UK doctorate at Palmer because fire recently destroyed the biology lab at the British base, Rothera, to the south. Bio is still how Palmer used to be, the ground floor a warren of labs and aquaria, a lobby with red float coats and full-length orange Mustang suits in three sizes hanging on hooks, stairs up to the galley, kitchen and food stores, the comms rooms and admin corridor, then up again to a central passage with ten small bedrooms, 'double berths' in Palmerspeak. Wooden cupboards and shelves along one wall, a pair of bunks along the other, the upper reached by climbing on the small desk. I find out who likes having their shower in the morning and get a spare slot: 6.45 a.m.

We assemble in the galley for our first station meeting, welcomed by Station Manager Bob Farrell. We eye each other. Some of the women have firmly pulled-down hats – the tradition of a winter head shave when Antarctic bases were all-male has jumped gender. Three years ago Bob was reliable, smiling Senior Logistics Person. Now he's our friendly 'Area Manager' with a palm pilot containing all the points he wants to make – to general mirth. Raytheon Polar Services, a subsidiary of the giant US builder of military weapons, currently has the contract to manage and provide support services for the three US Antarctic stations on behalf of the National Science Program. All support staff running the station are Raytheon employees. New titles have sprouted: last time's smiley housekeeper Alex is now Brenda the Administrative

Co-ordinator. Laid-back lab manager Rob is now Cara, Senior Assistant Supervisor, Laboratory Operations. The remodelled Palmer has beds for forty-six, but with departed scientists on the LTER cruise we are briefly down to thirty-seven: sixteen women, twenty-one men. Twenty-four staff are supporting nine scientists and four extras, including me.

Space is at a premium at Palmer. Raytheon has rationalised, and now support staff generally have their own offices. But Bill has pre-negotiated a desk for me, in T5. ('T' stands for temporary). T5 is up the track, a navy original on the edge of town, the last building before our 'Back Yard' and the glacier reaching across the promontory behind the station. Last time I used to scramble over the barren, heaped rocks of the moraine to loiter at the monster's very tip, listening to the tinkle of water melting under the shrunk, discoloured lip. When Palmer was built, thirty years ago, the glacier's snout ended a quarter of a mile away. Now it is a twenty-minute walk across the tumbled boulders of its leaving. The Back Yard lengthens every year.

After dinner I knock on the screen door of the Birders' tent, the ingeniously private work space for the seabird team, discreetly but permanently moored on a wooden platform outside Bio. The wire screen has a ragged hole, and instinctively I think – mosquitoes. I'm back in the southern hemisphere. But, of course, there are none. It's one of Antarctica's great gifts. Bill and Donna are inside catching up with their Field Team Leader, Chris. Chris, 6' 3" of lean muscle and few words, spends his summers at home in Alaska taking adventurers white-water rafting. He's the man to trust in boats, says Bill, and he hands me two small plastic ziploc bags. Inside are penguin ticks, *Isodes uriae* – big, grey, rounded, sluggish creatures engorged with penguin blood. Ticks climb up the splayed-out feathers of nesting Adélies and attach to their necks. Adélies don't groom each other. They are too aggressive. And they can't reach their own necks. The small active male ticks search for females on the ground among the penguin guano. The fertile eggs stay in the guano all winter until late spring, when the cycle can begin again. This season ticks appear to be particularly prevalent.

I think of penguins' fronts. Clean white feathers when they come out of the sea. Streaked with brown-pink guano after long hours of nest duty. But, now, stained with the bright red of fresh blood. The multifaceted nature of disaster.

Today, the day we arrive, one inch of new snow fell.

3

The life of a penguin

When does an Adélie penguin's life begin?

Is it at the moment of hatching – the sharp little hard white bump on top of the beak pushing through the eggshell, the two-day exertion of cracking the tough encasing walls that have protected the growing embryo, the culmination of the 31- to 33-day transformation of deep yellow yolk suspended in a thick white albumen to a 95-gram chick emerging into the cold air, barely able to hold its head up, covered in fine brown down, containing in its body sufficient yolk to sustain it for up to four days without food?

But hatching depends on successful incubation. Incubation depends on parents sharing nest duties, guarding up to two eggs and keeping them warm. The male, taking the first incubation shift, must have arrived at the nesting site having fed sufficiently well to fast for up to five weeks. The female, replenishing her reserves at sea after egg-laying, must be able to return to the nest in time to relieve the male. Both must continue to launch themselves into the ocean, find enough prey and, avoiding predators and all hazards of ice and water, return to land to sit out each incubation shift.

And hatching depends on the weather. The weather affects Adélies at sea, and on land: access to foraging areas, extent of sea ice, position of random icebergs, availability of prey, the amount and timing of snow, melt water chilling the eggs, winds repositioning snow and covering nests, ferocious storms sweeping eggs, or adults, away. And hatching depends on individual accidents and disasters – mini-avalanches, passing seals, hungry skuas, fights, death, and all the random mischances that can break vulnerable eggs.

Or does an Adélie penguin's life begin at the moment of successful fertilisation? The male balancing on the female's back, a coming together of the only rear-end orifice either has?

But track further back and the question rapidly unravels. Chick-raising demands a heavy commitment of resources from both parents. No more than two eggs are laid a season. Adélie penguins practise from a relatively young age the tough business ahead, visiting the colonies where they themselves hatched, checking out potential nest sites, briefly pairing, trying nest-building, possibly laying small, unfertilised eggs. In general, males are not physically ready for the commitment of breeding until the age of seven or eight, females (depending on the age of the male partner) younger, from around four.

Track forward. At hatching the chick needs warming, feeding, protection from predators. Even if both parents are of an appropriate age and experienced, the rigorous requirements of chick-rearing make heavy demands. If the chick is one of two, it must compete for food, reaching up, begging, treading on its fellow nest-sharer, grabbing. As it grows, it must get enough food and avoid being eaten, risk moving away from the nest, join other chicks in a crèche, chase after parents for meals, find its way to the beach or the ice edge. It must have gained a sufficiency of weight and condition at fledging. It must enter the sea, surviving the lottery of attending predators, swim for the first time, get under the surface, find and catch its first meal. Forage successfully. Manage the hazards of the first winter. Survive the first vulnerable year – the most risky for all wildlife – with potentially an 80 per cent chance of failure.

Because, in order for an Adélie penguin's life to begin, the stock of appropriate and experienced adults able to pair successfully and raise a chick needs to keep being replenished. Which means a sufficient number of Adélie penguins must survive as fledglings, and from fledglings to juveniles, from year one to two, two to three, each succeeding winter and summer, to reach maturity; and return as breeders from a winter season that has allowed sufficient caloric resources to be accumulated to manage a lengthy fast (for a male) and the requirements of egg production (for a female). Recruits must be added to the existing population of breeding adults, new breeders to replace those at the ends of their lives. Mortality and recruitment need to be in stable equilibrium.

Every year the defining status of a penguin population relates to events occurring before. What happened to each year's intake, the cohort? What was the fledging success rate? What was the survival rate of the fledglings during their first year? The poignancy of the time-lag, the full impact evident a biblical seven years on. The difficulties of unravelling cause and effect.

Penguins spend more time at sea than scientists gave thought to. Penguins' time out of the water is visible, observable. Scientists traditionally plan their research activities in Antarctica to coincide with the easier access and more benign conditions of spring and summer. The return of Adélie penguins to the land – this emotive migration, this journey from out of the ocean – was for most observers, and for decades of those who studied penguins, the absolute start of the penguin cycle. The land habits of penguins entranced. The absorbing, full-on, on-land lives of nesting breeding adults, the process of chick-raising, seemed to be all. The impact of the Antarctic spring and summer, animals arriving at breeding sites, the great blooms of phytoplankton triggered by light driving the food web, humpbacks swimming north at summer's end, contributed for decades to research attention being distracted from the rest of the year. What came to be understood as the high proportion of time spent by Adélies at sea was largely ignored. It was difficult to observe, let alone study. Research on predator–prey relationships focused on pressures on the summer food web: who was taking what. Sea ice and darkness continued to impose major constraints on research. Penguin biologists did not know where their subjects reappearing at the breeding sites had been, or what they had eaten while there. They could only guess where fledglings went for the first year of life, or where the juvenile birds lived. For a long time the assumption was that penguins 'went north'.

Parallels are everywhere. Elephant seals are generally described in terms of their ungainly bulky land presence, with their lives at sea, their dives – reaching depths of 1,200 metres – barely acknowledged.

But gradually the emphasis changed. Ice-breakers allowed some winter work in the ice. Banding, and long-term commitment to study sites, allowed data on populations to be built up over time. Advances in technology allowed satellite tracking of animals in the water: seals at first, then penguins. Data could be collected on depth, duration and location of dives. Ecological studies put emphasis on the totality of penguins' lives, and their physical contexts.

For Bill, at Palmer, the questions are clear. 'We have a good handle on what is happening locally. We know a great deal about what happens on land. What we need to know about is happening at sea. Where do the adult Adélies go in winter? What are they eating? Where are they finding it? What impacts on the availability of their prey? What is driving mortality at sea? The ocean – and, critically, the survival of the fledglings at sea – appears to be the key to understanding the Adélie. We can speculate about what happens. But this is the core of the issue. Penguins are seabirds. We only see a slice of their lives, a small fraction. When the fledglings leave, we lose touch, and we continue to be out of touch – with some glimpses at sea, during infrequent, expensive winter cruises – until they return as adults. Winter remains the great unknown. The gap in our understanding of how winter ecologies operate is large.'

Adélies and emperors are the only ice-dependent polar penguin species. The Antarctic landscape is vast. Adélies are patchily distributed. The Palmer islands are near the Adélies' northernmost range. And Adélies have no middle ground. They are either successful, or not.

4

Remnant Eden

Riding north from Rome in dust and heat, across land repeatedly tramped over and trampled by invading armies, to come, at last, to mighty Caprarola, palace of Cardinal Alessandro Farnese, nephew of the Pope. Entering the blessed coolness, windows shuttered, shallow bowls of melting ice on wooden tables. Mounting the helicoidal staircase, steps wide so the Cardinal could ascend on horseback, to the Great Hall of Maps. And on the end wall, facing down the length, the map of the world. Europe in central place, Asia and India occupying the east, North America to the west, the equator a bold yellow line dividing north from south, South America spilling below. The land narrowing to a strait sailed through by Magellan in 1520, then spreading out again to a wide triangle of confidently portrayed continent bisected by the Antarctic Circle, equipped with rivers and mountain ranges, and roaming large beasts. The world according to the latest thinking in this year, 1574.

Europeans exploring south from where they lived believed a great continent must exist to balance the land in the north. But as ships poked and swung ever further south, the ring encircling the globe below which land could be slipped lower and lower, a noose to catch an ever-shrinking continent. And when James Cook, pushing through mighty swells and increasing ice, managed between 1772 and 1774 to circumnavigate the globe further south than anyone had ever sailed, he concluded that if a continent existed it must be all or almost all below the Antarctic Circle, that imaginary line around the earth at latitude 66° 30' S, where for a measurable moment, in midwinter, the sun does not rise at all.

But Cook had found islands with that valuable resource, seals. English
and American ships were quick to follow. Agile fur seals hauled out on
secluded beaches for the late spring birth of their pups were slaughtered in their thousands, along with elephant seals, those ungainly
mammoths, heaving along over the stones. Fur to felt into warm hats,
blubber rendered down in cauldrons, to lubricate the machinery of
industrialisation.

Sealers searching further south for undisturbed beaches marked
routes and new coastlines on undisclosed charts. There was, after all,
something. A place of water, stone, snow and ice where the air was
clear and animals came to breed, and feed. A remnant garden of Eden,
undiscovered, unscarred, unexploited. Exploring expeditions came from
Russia, Britain, France, the United States. Captains read proclamations,
flagpoles jammed between rocks, shivering crew gazing astonished at
shelves of ice and distant mountain ranges. The post-lapsarian world
was impinging. Bit by bit, remnant Eden was becoming Antarctica.

The new land was inconveniently positioned for maps. The spike
holding the globe stuck through its potential, and the Mercator projec-
tion reduced the coast to a wavy line along the bottom of the accepted
world-view. But every geographical feature, every living thing, was
nameless. A cornucopia awaited. Iconic citizens, officials, friends, ships'
captains, home towns, expedition members were honoured, descrip-
tives – colour, size, shape – overindulged. A sea and a seal were called
Weddell after James Weddell, London-born master of a ten-gun brig
in the Napoleonic Wars, now sealer and explorer. A stretch of icy coast,
and a small bird about 70 centimetres high with a black head, a bright,
black eye ringed with white and a snowy white breast, were named
after a devoted wife waiting in France for her naval captain husband
to come home from exploring as far south 'as the ice permits'. And so
Adèle Dumont d'Urville was memorialised by a slice of Antarctica, and
untold millions of Adélie penguins.

The chance of timing dictated which monarchs were attached to discoveries. Englishman John Biscoe landed in 1832 on what he thought
was part of the Antarctic mainland and took formal possession on behalf

of King William IV, naming the highest mountain he could see Mount William. An ice-covered island further south had already been given the name of the king's German wife, Adelaide. Biscoe did his best for these brief incumbents of the British throne, but his mainland turned out to be an island, named Anvers after a Belgian province by its Belgian discoverer, Adrien de Gerlache, sixty-six years later. And in 1903–05 French explorer Dr Jean-Baptiste Charcot, making a rough chart of Anvers' south-east coast, discovered and named Mount Français, at 2,760 metres the clear winner over Mount William at 1,515 metres. A narrow bare promontory marking a sheltered bay was called Bonaparte Point, after the Prince, who was president of the Geographical Society of Paris.

Anvers Island, the intricacies of its coasts, its mountain ranges, remained largely unexplored until the British government established a base in Charcot's sheltered bay on the last day of February 1955. The British had been occupying assorted sites on the Antarctic Peninsula and nearby islands since 1944, for wartime and territorial reasons. Now a small wooden hut was built end-on to the north-west wind on a bouldery bit of ground just past Bonaparte Point, facing south across the bay. The tasks at Base N were mapping, geology and searching for minerals – in particular, copper. None of the six men allocated to Anvers could ski, or had any medical training, or knew how to cook. But they were young, on an adventure, managing with what they had – hobnailed boots not spikes, short skis not long – until the next supply ship arrived with whatever the authorities in London thought fit to send them. A magnetic researcher posted to one of the other British Antarctic bases insisted that several unusually shaped cases included in his scientific gear should be checked on the London dockside, before loading. Inside he found dentists' chairs. Five of them.

All men living in these remote and desperately isolated Antarctic outposts were tasked with some kind of scientific work. While there, it was believed, they should do something useful. Doing science had always been welded onto British exploration and territorial expansion. Science was a good thing. It occupied mind and body, imposed a discipline. It was acceptable public relations; it raised the profile of a project, helped with finding funding. And it produced tangible results: science meant things to bring back, proof of action – skins, skulls, fish, measurements, lists of figures, plants, rocks, fossils. Running through, and underneath, was the pleasure of pursuit, the seductive delights of discovery, the drive to collect. And there was also, without doubt, the passionate search for knowledge, the desire of scientists to find out, to understand. Peter Hooper, appointed commander of Base N as a 21-year-old geologist, reflected recently, as a retired professor: 'Our job was to map and survey. But science wasn't just a pretext to set up bases. The political rationale – occupation, making the British presence felt – did allow scientists who wanted to know more, an excuse for doing so.'

The bay in front of the hut buzzed with animal life for much of the year. Humpbacks swam off-shore, great mouths like coal scuttles scooping up quantities of small, shrimp-like krill. There were seals: Weddells, crabeaters, leopards, a few elephants. Even life wriggling in a freshwater pool, a 'job lot of beasties', preserved in a bottle of methylated spirits. Birds were everywhere: terns wheeling, gulls, skuas, sheathbills, hundreds of blue-eyed shags, snow petrels, Wilson's petrels, giant petrels most of the year. Penguins were always around, except for just a few months in winter: big gentoos with their splash of white above each eye seemed to arrive in pairs; chinstraps, sharp black line beneath their bills, hung out in groups. And of course there were Adélies.

The men tried eating some of the inhabitants (skuas tough, shags excellent). Hundreds of 'first eggs' were taken from Adélie nests because, according to accepted belief, an egg would be replaced. 'In any case, there were so few of us. And so many of them.' A small team of huskies supplied in the second year to pull the sledges needed to be provisioned locally with penguins collected from the islands, and slaughtered seals. The pups ran free until five months, disrupting nesting birds. Nothing could be done about it.

But it was what lay behind their backs that mattered. Anvers, 74 kilometres north-east to south-west, 42 kilometres north-west to south-east, was the largest and most southerly part of the Palmer Archipelago.

The men were here to survey, to tie the island into the complex coasts of the Gerlache Strait and the Antarctic Peninsula. Immediately behind the hut the ice rose in a smooth sweeping slope up to the top of the ice sheet covering Anvers, a steep climb. Man-hauling the loaded Nansen sledge to 1,600 feet where a depot was established, just under 4 miles inland, was a hard slog. And everything left here was always buried in heaping wet snow.

Through autumn storms and winter dark and on into the prolonged light of summer they made long reconnaissance and surveying sledge trips across the Marr Ice Piedmont, probing routes through the mountains, geologising whenever a rock stuck through the snow and ice. One man, usually the general assistant, Doug Litchfield, who had a bad back, remained at the hut, but Base N wasn't a met base and daily skeds (schedules) weren't required. Peter Hooper: 'We could do as we liked and report what we had done.' The mountains and glaciers on Anvers' eastern half were achieved, except for the impenetrable northeast corner. They climbed Mount Français, the highest peak in Antarctica north of the Antarctic Circle, and hit the London headlines; and they did it three times, because cloud kept denying photographs from the summit for surveying purposes. In their small dinghy, with its single outboard engine and no replacement parts, they criss-crossed the sea to Base A, Port Lockroy, just off Wiencke Island – perilous, risky journeys. But the western half of Anvers, the weather side, defeated them. To their right the mountains rose like a boundary wall. Ahead and to the left the ice plateau was constantly mist-shrouded, swept with gales, snow-thick. The ice surface sloped away into the mist and murk, edges dangerously hidden and crevassed.

No one at Base N was a biologist or zoologist, but with birds so 'tame' and wildlife so abundant, everyone was encouraged to note observations of local wildlife in the biology notebook, a Letts diary kept on the bench under the window in the base commander's office, with the rock specimens and the precious microscope. Preliminary printed pages with dates for the Oxford and Cambridge terms, times of high water at London Bridge, an advertisement for 'Figure Studies' (prints offered to artists and sculptors), then, inside, random but careful dated notes by interested amateurs. Work was done on the local Palmer seabirds. Numbered rings were attached to giant petrel chicks and skuas, and details noted. The arrival of vast numbers of Adélies to breed on the islands every October was recorded, times of hatching observed, nests counted. On 23 December 1955 'a very rough count' estimated approximately a thousand occupied nests on the east point of Big Todday (later renamed Litchfield), 'over three thousand' on Little Todday (now Humble) and 'from eight to ten thousand nests' on Penguin Island (now Torgersen). As far as anyone could tell, none of the many gentoos or chinstraps nested locally. A sketch map was made of the area – including Arthur Harbour, named after the tall, daunting governor of the Falkland Islands – showing the distribution of bird colonies colour-coded for Antarctic terns, skuas, gulls, giant petrels and Adélies. 'This is an obvious place for a study of Adélie penguins', the diary noted firmly. And, significantly, 'We understand that the sea ice conditions influences the movements of birds and mammals.'

But British government officials considered that studying penguins conferred little economic or scientific value. Biology was the designated science at only one British base. Counting penguin nests and attaching numbered rings to the legs of birds was incidental, spare-time stuff. Long-term data sets were not significant. Penguin studies, when they did occur, tended to happen by default, and to be carried out by non-biologists. The three species of pygoscelid (i.e., rump-legged) penguins – chinstraps, gentoos and Adélies – nested on open, pebbly ground, ice-free, with access to the sea, the kind of rarely available space explorers and scientists needed to set up camp and build huts. Penguins, as neighbours, were easy to observe. To lonely men they provided endless entertainment and opportunity for humour: obvious candidates for surrogate humans, penguins wore evening dress, walked upright on two feet, lived in families and congregated in large groups. Their short, flapping flippers were inefficient, lacking fingers, but still, enough like arms. Walking with a rolling gait, they were clumsy, argumentative, noisy, dirty. They fought. They stole. Penguins, particularly Adélies, became that infinitely patronisable combination, low-class locals in upper-class kit.

In 1911 Murray Levick, Royal Navy surgeon with nothing much to do while at Cape Adare, at the northern entrance to the Ross Sea, made one of the first studies of Adélie penguins. Thomas Bagshawe, a twenty-year-old geologist, and M. C. Lester, a slightly older second mate from a tramp steamer, were marooned (intentionally) in 1921 at Paradise Bay on the mainland opposite Anvers, and obsessively studied the local gentoos. In November 1948 medical officer Bill Sladen focused doggedly on the neighbouring Adélie colony at Hope Bay, at the top end of the Antarctic Peninsula, when fire killed his two companions and destroyed the Base D hut. Alone, in shock, he began a detailed study of the Adélies' breeding biology, dissecting birds killed to feed his dog team.

Three years after work began at Anvers, the British government announced that surveying and geology were complete. The Duke of Edinburgh and his entourage, including the American observer of the British bases, Crawford Brooks, who always carried tennis racquets and balls ready for any chance to play tennis on the snow, called in to Base N on a royal visit in early January 1957. By December the base was closed, the hut locked. Science in Antarctica was changing, becoming formalised. The International Geophysical Year begun in 1957 had put an emphasis on collaborative work in inaccessible places, such as space and Antarctica. Twelve countries operated forty stations on the continent, twenty on sub-Antarctic islands, with more than 5,000 personnel.

At the end of the IGY the incentive to continue long-term science projects initiated in Antarctica was obvious. Data was being pooled, programmes co-ordinated. Despite political rigidity, in spite of Cold War tensions and propaganda, international scientific co-operation was working. But the threat of international conflict spreading into Antarctica was real. Nations competed openly. An alphabet of Argentina, Britain and Chile triple-claimed the Antarctic Peninsula. The USA and USSR made no formal claims but reserved the right to claim. Science, it seemed, might be the means of providing a pragmatic solution to a difficult problem. Science could become the currency of the region's politics.

In Washington DC small groups of scientists and diplomats began brokering a treaty. The terms were kept deliberately simple: complexity could have tripped the delicate negotiation process. Signed on 1 December 1959, coming into force on 23 June 1961, the Antarctic Treaty was a cunning balance, a compromise. It side-stepped. It achieved objectives by avoiding confrontation between governments with strong claims, weak claims, and non-claimants. The long momentum of territory taking in Antarctica was halted. All claims, it was agreed, would be 'set aside'. Antarctica was placed inside a cordon of simple rules. It was to be used exclusively for peaceful purposes. The compromise, the glue – the raison d'etre for all human activity in Antarctica – was to be science. Scientists were to be the continent's guardians. From now on human endeavour in Antarctica, however structured or motivated, whether transparent or holding manoeuvres, an integral part of a government's policy or keeping the lights on, whether parochial, or routine, or large-scale projects, would operate within the context of the doing of science. Generally unheralded, almost unnoticed, a way had been found to commence protecting this last-discovered continent. Belonging to no one, with no permanent inhabitants, no government, no police, Antarctica's chief export was to be scientific knowledge.

Scientific stations could be set up anywhere by signatory nations. The United States had already placed a base at the south pole, Amundsen–Scott South Pole Station, built on top of 2,740 metres of ice, and another at McMurdo, colonising the myth-laden tip of Ross Island, where Shackleton and Scott had set out on their journeys towards the south pole. Early in 1963, after surveying thirty-three potential sites for their advantages, a small team selected the deep anchorage of Arthur Harbour at Anvers as the site of a third permanent US base, named Palmer Station after the American contender for first sighting of the Antarctic Peninsula, the 21-year-old Connecticut sealing captain Nathaniel Palmer. The new station was to focus on biological sciences. In

January 1965 a prefabricated T5 hut was put up next to the British Base N hut, which was unlocked and lent as a biology laboratory for the use of American scientists. Two years later work began on a larger station on Gamage Point, a stubby promontory just across the inlet from Bonaparte Point; it was finished in March 1970.

The definitive outline of Antarctica had by now been revealed: thousands of newly created satellite images, glimpses, cloud-obscured, pieced together to create a startling picture of a naked, improbably cloud-free continent. A sea-ringed stingray heading across the ocean towards the gap between Australia and South America, truncated tail kicking up to South America.

Pinned to my study wall is the latest map of Anvers Island, matched and patched from satellite images. It reveals an elegant white fractal shape in a blue-black sea, schizophrenic halves stitched roughly together. The complex, indented east coast, ridged and spined with mountains, deeply shadowed, is like a Rorschach ink-blot test. The western half is curiously unsettling, the interface of the ice sheet with the sea unfocused, the surface appearing soft, even-textured, indefinable. Ice encompasses the entire island except for tiny scratchings, minute short rootlets of brown, scattered scraps along the southern coast: the small islands and low peninsulas of the Palmer Station area.

But for us, living at Palmer – we can only sense the island at our backs. The limitations on our movements are absolute. The men at Base N explored their local world. We experience only a fraction of the whole: our few fragments, our brown squiggles, and the sea between, on Anvers' southern coast.

5

Seeing for myself

Tuesday 8 January 2002

The sun shone day after day, my first summer at Palmer, and the Adélies panted in temperatures around 0°C. The ground was pinky brown, stained with the guano of thousands of penguin meals, the discarded remnants of hundreds of thousands of krill and small fish, each individually located in the ocean, caught in a penguin's beak, swallowed and transported back onto the land. Parents ran up from the water clean and shiny, and the stones on the ground clinked as they passed. At the nests the relieved partner left, dirty but duty done, pulling its body in tall and thin, flippers tucked close, weaving and swerving through the gauntlet of aggressive nest-sitters like a rugby player trying to avoid tackles.

There's a revelling in the intense activity of a penguin summer. Its rhythm catches you up. It's there in all the accounts – the early explorers, scientists, delighted visitors, dedicated penguin observers, everyone engrossed in the privilege of watching, by the luck of being there. Becoming in a small way part of it, because they are tolerated. Permitted. Stop watching, and you miss something. Keep watching and you begin to recognise the stages.

But this time.

There is so little noise. So little smell. Such small groups. So few chicks. An almost complete absence of guano, that starburst of pink radiating out from each nest, that signal of occupancy, of chicks at home, of regular feeding, of the need to feed, of rotation of parents with their full bellies coming back from the ocean. Some of the smaller colonies have only one successful nest with one chick, very occasionally two, under the one bird. Seeing the Adélies for myself is shocking.

In my head are memories of busy, functioning penguin colonies. The din of living, the pervasive smell of food being crammed in and processed out. Of beaks snapping and clashing, of the haze of dust and feathers rising over massed nests. By the end of the first week in January 1999 nests were beginning to lose their discrete shapes, be trodden down, the carefully accumulated nest pebbles scattered. Woolly grey chicks were starting to wander. Some clustered in mini-crèches, long flippers hanging like oars, feet too big, like clowns' feet. Single penguins patrolled the edges of colonies, facing outwards, watchful, cocky. Skuas swooped, or strutted, bold, looking for opportunity. The cackling, calling, grunting, the insistent cheepings of chicks, the strong distinctive smell, filled the air.

Now, at the same point, the end of the first week of January, most chicks are still very young, helpless, lying on their bellies. The colony outriders – those outward-facing singles – are hardly in evidence. Now there are so many singles, or penguins just standing around in twos and threes, it is difficult to understand if pairs really exist. Difficult to see if a nest is still in place or scattered, as in the crèching phase. There is less aggression. Less need for a penguin to run the gauntlet to get to its nest. Less pecking, and flapping. Quieter, not hectic. Subdued.

There seems a loss of structure. Colonies do not appear to be acting as a unit. There isn't a sense of a society engaged in group activity. Last time each colony, each subset, seemed to me like a suburb, most households roughly similar. Now the rookery feels like an urban city in a war zone, with huge differences between the households. Some colonies are reasonably active, some almost non-functioning. But in general the city is severely depleted. There appear to be very few 'families', lots of singles and childless partners. I want the impossible. To have a map of the streets and houses, so I can see what has happened to each household compared with before. I want to be that detailed. What has been the history of this couple? That nest site? What are the dynamics of success? Why did this nest succeed, that fail? How has disaster affected each one? It feels very personal. Each has its own story. I want to document it that closely. Afterwards, UK penguin biologist Bernard Stonehouse gave me a context for my emotions. 'Two to four thousand is a "nice little colony" – the right kind of number to study. Too big and it's difficult to manage. Too small, and you get involved in the fates of the individual birds.'

One leopard seal has been working the area periodically, another full-time. The pickings are easy at Torgersen, where birds have to stack in bottlenecks to come ashore. Beach access has been confined by snow to two narrow locations, and the water churns as a leopard thrashes a penguin out of its skin. Birds fly over, grabbing morsels. If the dead penguin is one of a functioning pair – this season that's not just a loss, it's a disaster.

There's a small amount of pebble-carrying and nest-tidying, but very little. I see one pair attempting a fumbling copulation: beaks clacking, flippers waving, male attempting to balance on the female's back. Many birds are sitting in the brooding position. But nothing is happening. What do birds do when the eggs have failed? Does the pair bonding remain? Does alternate feeding continue when there's no need to relieve each other on the nest? I find just one empty egg on a rock; but no eggshells. I see several dead penguins on the ground, bones and sinew but the carcasses could belong to last year, or the year before. The skuas seem particularly confident.

Where have all the penguins gone?

Hours after dinner I go into the Birders' tent. Bill, Donna and Chris are sitting at the high wooden worktops among keyboards and screens, empty coffee cups, reference books and photographs of Donna's much missed dogs and horses in Montana. Heaps of essentials lie ready to hand or in beer crates of gear: coats, hats, gloves, float coats, sea boots, backpacks. They list their day's actions. Visits to the five inner-island study sites counting totals of attempted breeders. Checking snow depths on three islands, and numbers of brown skuas and of giant petrels on two. Monitoring selected Adélie reproductive sites on Torgersen and Humble. The season is in disarray. Data collection is behind schedule. For Bill so many factors to be juggled and thought through.

Last season the weather was ideal: average quantities of snow, excellent working conditions, only five days of fieldwork missed through weather. Results were as expected.

This year is the poorest Adélie reproductive success rate ever measured at Palmer. That much is obvious. But something else has happened, totally unexpected. Bill: The number of penguins that actually showed up and tried to breed was down 50 per cent from last year.

Why? Did ice slow their journeys? Or was it something else?

Bill doesn't want to commit. 'I'm not able to figure it out for a while. Things are still fluid. It could be krill are hard to find. There seem to be no whales feeding. Humpbacks need krill.'

Next day Bill explains. They were bone weary, all they wanted was to eat and drink. They couldn't formulate their ideas, couldn't sound intelligent.

I need input. I need to find out what happened before we got here: what went wrong.

6

A field season from Hell

August to December 2001

The sea ice came late on the western side of the Antarctic Peninsula in winter 2001, then grew, fast. In August scientists on the powerful US research vessel *Nathaniel B. Palmer*, working south of Adelaide Island in Marguerite Bay, on programmes of winter research, reckoned they were seeing a good ice year. Then, late in September, persistent gales began blowing, bringing thick, soft snow. A strong low-pressure system settled around the top of the Antarctic Peninsula, coupled with a blocking high in the South Atlantic, dragging warm, moist air down from the north. A blocking high is an anticyclone that remains quasi-stationary for several days in the mid-latitudes. This time the low backed by the blocking high did not go. It stayed. The winds blew the sea's covering of ice against the peninsula's western coast, compacting the floes, rafting, piling. Metre-thick floes of first-year sea ice became 15-metre-thick accretions – clumping, baggy, ballooning, frozen structures hanging deep in the ocean. Blizzards dropped thick, soft snow on top of the sea ice, heaping it on top of the snow already there. The ice caught the *Palmer*, trapping it at 68° 16' S.

At Palmer Station, 385 kilometres to the north, the sea started freezing at the end of September. That fast, extraordinary stiffening of the shifting waters, that stilling of the swells, that flattening of the unflattenable which can happen at Palmer any time from midwinter on, happens. The ever-moving, heaving, rising and falling is suddenly motionless. The sea can be grey, black, green, violet, blue. It can sheen and shimmer as if its vast surface is coated in silk. It reflects and hides. Now the sea is white. Sparkling, solid white. The ice encompasses the near islands, linking Humble and Torgersen, tying Litchfield to the station buildings on Gamage Point in ruler-flat surfaces. What could not be walked on can now be walked on. The sea ice may be only 20 centimetres thick. That is enough. Islands can be reached on foot.

The year at Palmer had been drier than normal, with August snow-fall the second lowest of the decade. Now the early spring storms are bringing heavy snow, with significant drifting.

Adélie penguins must use their run of summer time efficiently. On midsummer day, 21 December, the longest day, Palmer has nineteen hours of daylight. Two months before the longest day the Adélies are starting to pair. Two months after, the new chicks should have left the islands. Males begin arriving at their Palmer nest sites in early to mid-October, two to three days ahead of the females, well fed, primed for breeding, plump with energy reserves that must last them till late November, when they can leave their first brooding stint and feed again at sea. Thousands of Adélies come swimming in, walking in long lines one behind the other over any intervening sea ice, or tobogganing on their bellies, feet pushing, flippers paddling.

Except, this October, most nest sites are under snow. Penguins face long hikes across the ice and arduous climbs up corniced snow banks over-hanging the beach rocks. Snow drifts hide the favoured slopes, the hollows between rock slabs, the ledges used the previous season and the season before that. The small stones needed for nest-building are inaccessible.

Cyclones tracking between South America and the peninsula keep pounding in through October, bringing high winds, storms and snow, with 50 per cent more precipitation than the next-highest month of the previous decade. A massive storm on the 28th drops more snow in one day than an average month at Palmer, burying everything. Litchfield Island looks like a well-iced cake, scalloped in two-metre deep drifts, a few rocks standing up in thin, flat-sided spikes, the steep hills rising like dwarf alpine peaks. None of the contours of the nesting colonies, none of the ridges and high places, is visible.

Disoriented, delaying but driven, penguins begin pairing and nest-building.

The Adélie penguin work, the central focus of the seabird ecologists, begins at Palmer with the arrival of adults to claim nest sites and ends with the departure of the season's product, the fledglings. This season the annual Adélie breeding biology studies are due to start on 15 October. Two field assistants have been booked to begin the data collection. But field team leader Chris and eager, hard-working Heidi are trapped on the *Palmer* in Marguerite Bay, in rafted, shattered ice compacted by the unshifting winds. Sixteen thousand kilometres north, in Montana, the remote valley where Bill and Donna live is too isolated for television or mobile phones, but they can receive the satellite phone service iridium. They persuade Brett, Palmer's winter assistant lab manager, to leave his job and become an instant fieldworker. Brett is tall, solid, even-tempered. He has a research interest in parasites, a passion to reach every continent before he turns thirty and a sideline business in locating and selling secondhand off-road vehicles. In Antarctica he's unloaded the big C130 aircraft at America's south pole station but never worked with penguins. Now he puts on snow shoes and tramps across the sea ice to begin the season's penguin counts. Only six days late.

The state of the ice at Palmer is the boating co-ordinator's responsibility. Jeff, from Wisconsin, has only just got the job. There are no ice augers, so he saws holes through the top ten inches of slushy snow, through solid ice, to the water. Jeff: 'The sheer excitement of walking on the frozen sea. Then the shock of seeing the water at the bottom of the hole, and being forcibly reminded we were standing on the ocean.'

In Montana, Donna creates digital images of the study site penguin colonies and emails them to Palmer. But in the confusing white smoothness penguin colonies can seem mythical. Distinguishing marks are obliterated, snow stakes for measuring depth are buried. Only a few piles of bedrock stick up, with sparse penguins on the crags like lonely Roman sentinels. Brett phones north to Montana: 'I'm standing here. Now where?' He phones south, to Marguerite Bay. Chris and Heidi try to explain from the maps in their heads: find this rock, walk so many paces. But it's a struggle. Usually by late October most of the birds that are going to arrive at the Palmer nesting sites have done so. But Brett can find very few birds. The colonies are very thin.

Chris and Heidi finally get through to Palmer three weeks late, releasing Brett for an overdue break north, to the joys of cities and anonymity. Early summer seabird work schedules have slipped significantly. There are data to collect for species-specific time series, protocols to meet. Skiing out to the colonies, they find a few birds with a scatter of nest-building stones. One, in the absence of anything else, has brought its mate the flipper of a dead penguin. Almost every bird is sitting in a hole in the snow created by its own body warmth. This time last year the Adélies had begun laying. Now birds are still straggling in across the ice. Chris and Heidi don't know where the birds are in their breeding chronology.

A wild storm on 10 November starts the sea ice cracking apart. Next night winds peak at 76 knots, and by the 12th the ice breaks rapidly, well-beaten trails to the islands sailing off at speed. The following day pieces of ice festooned with bamboo route markers and red and green flags float back in. The sea is a mass of brash and broken ice, too choked for most zodiac work. Surly squalls and high-wind storms deny boating. For Chris 'the office' is the islands, Palmer is back home. There's everywhere to go, and no way to get there.

The Adélies have delayed mating. But in this dismal situation, mid-November, eggs begin to be laid. Many are put down directly on the snow, sinking straight in, the second egg following on top of the buried first egg. More storms hit, heavy squalls and fierce winds rolling in from the north-east, dumping snow in big, loose flakes that collect on the sides of the penguins, blinding the birds, burying them to their beaks. Then winds switch to the south-west, repositioning the snow, loading it in totally different areas. Birds that had been sheltered from the storm systems now experience their force. Re-depositing drift snow hits every nest but, most of all, those in lower colonies, off the ridges.

Chris: Even getting out of the zodiac and tying up involves leaping ashore with an ice axe, and the multiple forms of snow anchors usually used on mountaineering expeditions. He and Heidi trudge over the islands in snow-shoes. They worry about walking on unseen birds. Nests are burrows, or humps under the snow. Some of the colonies are completely buried. Heidi: 'Make a little noise and they stick their beaks up, through small peek holes. You look through and see upturned black beaks, the staring eyes, of birds sitting tight down a hole on their eggs.' Then next day another 10–12 centimetres of snow falls, covering the living birds beneath.

The weather continues to crunch through November and into December, one heavy storm raging in after another, visible on the Station TerraScan satellite imaging system as a massive twisting hook bearing down, bringing thick wet snow. Then the wind switches, south-west, re-piling the drift. Over and over again Chris and Heidi can't get out, or they are forced to retreat. But quick visits are managed to the furthest sites, Dream Island and Biscoe Point, to monitor Adélie numbers and count populations of gentoo and chinstrap penguins.

In the penguin colonies brief changes in temperature bring enough thaw for frantic birds to start mining rocks for their nests and begin taking from each other, desperate to get stones underneath the eggs. Then skies turn solid and dark, the squalls and flurries belt in with more snow, more drifting. If birds stand up to shift position, snow falls on the eggs, to melt into a puddle. Eggs are crushed or kicked out of nests as birds try to deal with the snow. Or the eggs lie cold, flooded out.

Chris, working at Torgersen, 3 December: 'It will amaze me if anything lives out of Colony 4 this year. 6 is buried. The few adults left are at least eight inches to one foot below daylight. The 7 complex isn't as bad, but they're taking their hits as everyone. 23 is doing poorly, but 20 is by far the worst. It seems like a lot of birds are pulling the plug and calling it a season. No breaks for these birds. The weather has been relentless.'

Any period of melting reveals a mess. Penguins appear from under the snow, and mass abandonment begins. Empty foxholes are everywhere, most with a two-egg clutch lying in snow or melt water. As always, the energy-depleted Adélie females leave the males brooding the eggs on first shift duty. Returning, they are finding the males already gone. Or the females are not returning at all, and starving males abandon the nests. For two weeks brown skuas have been circling over the colonies unable to harvest eggs under the snow. Now they feast, dining at the nests, at their leisure. Snow banks are shrinking, but storms still thrash in, with rain beginning to alternate with the snow. The weather is so continuously extreme, so demanding; 'Nothing crazy to report', the field notes record on a rare 'normal' day.

But some eggs do survive. Birds are hunkering down in the colonies, the vast majority either failed breeders or birds scouting for next year, birds that came and left. Chicks begin hatching towards the end of the third week of December: first one small, grey, fluffy, black-headed scrap, legs splayed out behind, then the second, the reserve chick. Nests are touching dirt, at last. But lows keep spinning in with heavy squalls of snow or stinging rain, kicking up white caps on lumpy swells. Chris: 'Standard misery for the Adélies.'

Heidi takes a hundred digital pictures a day to help Donna with data. Minimal opportunistic baseline breeding and population data on the other seabird populations in the area are achieved. Most of the work on south polar skuas is dropped: the birds are a month late, and they can only find four nests. Their zodiac rocks so much they can't tell if the kelp gulls are present or not. The giant petrel rounds have to be put on the back burner, but brown skua work is achieved. Ticks are collected from infested Adélies, as requested by Bill.

Several of the Adélie colonies become total losses; others are just barely holding on. Chris: 'We are recording one big mess. Failures aplenty.' Conditions from island to island are varying greatly in amount of snow, timing of Adélies' egg-laying and losses. Litchfield is still thickly snow-covered and seems the worst hit; the number of active nests keeps plummeting. By Christmas the biggest, Colony 8, has forty nests left. But it is difficult to tell how many of the birds on the nests are returners, just sitting it out – 'fakers' in Birder language. On Humble Island elephant seals arriving at the islands from the north plough through the already depleted Colony 2, smashing chicks.

But by the last day of the year all eggs have either been hatched, lost or addled. All checking under birds is over.

Parent birds struggle on with the routine of chick-rearing: one sitting on the nest, one finding food in the sea and bringing it back, greeting the thirsty, hungry, grubby partner, then depositing lumps of krill into the opened beak of the frantic biggest chick as it tramples and pushes aside its sibling. Then into the beak of the smaller chick, if it has been sufficiently persistent. If a second chick exists. This season one-chick nests far outnumber two-chick nests. This is the real story of the sorry season. There are not many chicks.

Heidi: 'It was a field season from hell.' Over an eight-week period storms or ice denied fieldwork two weeks and delayed or curtailed another nine days. 'Normal' weather on the Western Antarctic Peninsula brings prevailing winds from the west, and rapid changes. But now the winds blew from the north-northwest across the peninsula in a broad band. Mild, loaded with moisture, strong. Unprecedented quantities of snow and rain fell. Surface air temperatures rose.

On Station, boardwalks had been U-shaped paths between 8-feet-high snow banks. Snow obliterated stairs, and hid the multiple cables and pipes, the service lines carrying the Station's waste, water, fuel, power and communications that run like mini-multicoloured commuter lines above the ground. People built snow caves. Some got chickenpox. Everyone worried about the buried penguins. Everyone shovelled snow, every day, wishing it would go away, thinking it would get better while it got worse. At the end of December there are still big snow drifts on the islands.

Chris and Heidi worked in the Birders' office, the double-lined tent. A soft greeny-blue light diffused through the tent walls, coloured mini-bulbs looped across the ceiling. The tent was a precious commodity, privacy behind canvas, a single door that could be kept shut. Chris staked out his territory, Heidi hers.

Now they rush to get it ready for Bill and Donna. 'We had to shovel that place out.'

7

Penguin pebbles

Bill hands me two small stones with a seductive dark patina, curiously silky to touch. One wet March day in 1990 on Cormorant Island the rain had washed all the penguin guano off the surfaces. Bill noticed that some stones were gleaming but not others. Why? And why here on Cormorant, and none on Litchfield Island, where he had just been? There the stones were coarser, unpolished. Here he saw that the smooth dark stones formed narrow paths, tracks leading to current nests. But they also revealed routes to abandoned colonies. Every stone had been polished by the feet of penguins – thousands and thousands of penguins passing over them, year after year after year. The implication was a longer period of use on Cormorant Island than on Litchfield.

Another March day: 1997. Bill was kneeling by a square pit on a ridge at Biscoe Point while palaeobiologist Steve Emslie removed the layers of an abandoned Adélie colony, sifting through decades of use down to the underlying bedrock. A heap of pebbles lay on the ground, mixed with old, dark-stained penguin guano, littered with bits of bone, fragments of eggshell, feathers, fish scales, squid beaks. Head bent, shoulders hunkered against the wind, Bill heard a sound. The single honk of an Adélie penguin. He looked up. Surrounding the hole, the heap, him and Steve was a circle of Adélie penguins, edging closer, then stopping. Heads cocked on one side, staring out of one eye. A few steps closer. Stopping, staring out of the other eye.

Typical Adélies, says Bill. They'll come up really close and look hard. Then go back. They seem to want to know what is going on. They're curious.

Radiocarbon analysis of the excellently preserved organic remains from a range of sites was revealing that Adélie penguins have nested in the Palmer Station area for at least six hundred and fifty years. They were here when Ferdinand Magellan's three vessels pushed through the complex channels of Patagonia and found a route from the Atlantic to the Pacific in 1520, and when the Black Death slunk in through English seaports in 1347, sweeping one in four people to their grave. They were arriving along this southern coast of Anvers Island when persistent rains in northern Europe rotted crops in the ground, year after year, turning the fields to mud and villagers to gaunt, famished skeletons. Analysis also revealed that the only penguin species to breed here on the Palmer islands during the last seven centuries has been the Adélie; except for the last few decades, when chinstraps began establishing nests.

Palmer's landscape is small-scale. Land meets sea in a complexity of promontories, inlets and coves. This is drowned landscape, land half-revealed, half-submerged. There is a sense of transience in the boundaries. An ice sheet once hid all these contours under a frozen mass. As the ice retreated, islands happened according to the dictates of sea-level, and of glacioisostatic uplift – land slowly rising, no longer bearing its heavy burden. Some islands have been visibly smaller. Pebble banks ring their waists, indicating old beaches.

In the immensity of Antarctica small islands have real power to engage. Each of Palmer's islands, rocky scraps poking up out of the sea, looks different. Some have moss beds, brilliant green billows of hundred-year-old growing, miniature Tolkienesque forests inhabited by scurrying mites and transparent rotifers. Some have curious straight-edged, pebble-strewn surfaces marching briefly across them like truncated Roman roads. Some have steep-sided hills and sharp-pointed slices of crest rock. Brief patches of Antarctica's only two native vascular plants grow thumb-high in protected hollows, tufts of shallow-rooted hair grass, *Deschampsia antarctica*, and discrete compact cushions of Antarctic pearlwort, *Colobanthus quitensis*. There are eddies of a kind of fine mineral soil, and unexpected mirror-surfaced summer rock pools with precious fresh water. Everywhere lichens colonise the rock faces with bold splashes of Impressionist orange or flat white discs like moulds in a Petri dish, or fringes of desiccated black. In other parts of the world rocks are bedded in earth, pinned down by tree roots, netted by creepers. Here boulders balance precariously. The surfaces are fractured, scattered with shards and fragments, visible victims of cold. This is stripped landscape, the body laid bare. My mind's eye is inhabited by the ancient worn rocks of Australia. But to North Americans, this landscape is familiar like so much else here. Managing loose rocks, ice, snow, is second nature. 'Put your foot on the apex,' advises Bill, as we work across the rocky terrain. 'Don't go for the smooth surfaces. Balance on the points.'

Evidence of glaciation lies all around. Wandering in the Back Yard, I can peer at the hard sullen rocks, the heaps of detritus from the visibly retreating glacier, looking for dots of moss, small scabs of lichen. This land so long hidden is being re-colonised; the process of living has begun. But charting the rate of re-colonising has not been a project at Palmer, nor measuring the speed of the glacier's retreat on our doorstep. Research budgets are already allocated; these have not been priorities.

There is a tension between the brief opportunity of the moment and the time-scales required for understanding, between the geographic confines of the study area and the significance of the physical context beyond. Science in Antarctica has by definition been 'first work', stamp collecting. The Europeans' desire to discover for themselves a new world which in truth had always been there, but not known. Physically challenging, harsh, filled with unknowns and, perhaps most seductive, untouched by our persistent, pervasive species. A vast, ancient continent, not hung around with pre-existing descriptions, belief systems, sacred sites, with histories of conquest and oppression, with the chiselling-out of resources, with manipulated landscapes.

From the beginning the techniques of exploring Antarctica and what lay within and around it were a mix of survival skills learned from northern polar people, accepted familiar practice and leapfrogging, trying out the latest technologies. At Palmer the techniques of

the seabird ecologists are a mix of nineteenth-century and latest technology. They are human-intensive: footslogging and boating raids in
wild weather, counting by eye and measuring by ruler and hand-held
scale, rock scrambling and painstaking analysis of diet samples. And
they are cutting-edge: computers, mathematical modelling, satellite
telemetry Each season data and observations are recorded according
to established protocols. The time of looking is brief, in the order of
things. But in terms of Antarctic science the seabird data at Palmer are
relatively long.

Now, suddenly, Bill is working on a large scale, trying to understand the relationship between demography and climate change. 'Why
are Adélie penguins here on the Antarctic Peninsula? Populations
increase, and decrease. They have their rhythms. At Palmer we have
two cut-off points. Around seven hundred years ago the environment
became favourable for Adélies. What circumstances meant that islands
at Palmer became suitable? Now it is becoming less favourable.'

And he describes a day in February 1995, working on Humble
Island, 'doing what we had been doing for years. In penguin guano up
to our knees. Catching fledging chicks and weighing them. Monoto-
nous, dirty, miserable work. The worst working conditions. Covered in
penguin shit. Working like maniacs, getting the numbers we needed
for our study.'

As the numbers developed, Bill looked at the weights All the chicks
were very light compared with past years. 'A light bulb went on in my
head. I suddenly realised that I could be looking at a mechanism by
which the demography of birds could be affected by climate change.'

Reaching over to his laptop, Bill pulls down a diagram of Adélie
chick weights at fledging, an accumulation of fourteen years' measuring. The drop in chick weights for 1995 is clear. The 1994–5 season
experienced heavy snow. The chicks had been hatched late, as has happened this season.

'Lots of ideas come about because you are doing something you've
always done. You have a feel for how things should be going. And for
something different. Being down here so long, I have a feel for the

anomalies. When something does not fit the picture, it's very, very revealing. The progression of ideas has depended on being here, in the Antarctic, for much of my working life: seeing at first hand how the environment changes.'

The irony is sharp. 'We have done research on what disturbs a penguin locally. Tourism. Habituation. The impact of scientific work. Cause and effect. And now environmental factors have arrived bigger than us all. Suddenly we are trying to understand the big questions. How do ecosystems respond to climate change? How is the system here changing? Complex large-scale forces are at work. The issue has arrived at our front door.'

8

Palmer Day

Thursday 10 January 2002

Thursday morning, 1 a.m. Maggie the diver, gym-slip thin, sleeps tight as an ammonite in her down sleeping-bag on the wooden boards of the boatshed veranda. The sea sucks and slaps at the rocks just below, cajoling rhythms, merciful sounds undrowned by the whining beats of the Station generators, the thrums of refrigerator motors and air-conditioning units. At the boat tie-off the zodiacs tug and agitate on their mooring lines. Angular shards of ice jostle in the bay, herded as though inside an invisible rope by wind and waves. Dave, the power plant mechanic, checks the generator building. He knows by smell and sound if anything isn't quite right. He's a five-hours-sleep-a-night man.

4 a.m. A pair of sheathbills run along the roof above my head on their unwebbed grey feet, clatter clatter. Stop. Stare, clatter back. And repeat. Chicken-size, white-feathered scavengers of penguin guano and broken eggs, they sidle around Station pecking and poking. Last season a pair nested in a burrow under the trash incinerator by the mash and grind, taking the shiny incinerator key with them. We have to put our tool-boxes on top of our gloves, say the trade guys. 'The flying chickens steal the gloves. They don't want the tools.' By the zodiacs a leopard seal surfaces, mouth gaping wide to bite a port-side pointy end, then drops back down through the clear water, sleek and spotted, foiled by a rubber guard.

5 a.m. Maggie crosses the bare tossed boulders of the Back Yard and starts up the glacier on the permitted track. Small tents tucked in sheltered gullies hold people who choose to sleep out. The only rule: name signed on the Station blackboard, cancelled on return. A Palmer *habituée*,

she extends the physical boundaries of her time here, carving privileges out of knowledge and ability. In the galley Jennifer, this season's early Dawn, gazes at the rock ridge of Bonaparte Point across Hero Inlet as she bakes bread, logging the changing light, the contours and colours of this precise small piece of landscape in her mind. A teacher and dog-sled driver, she's on sabbatical from her high school in Alaska.

5.30. Maggie comes into the galley from the glacier. 'You get used to people at certain times in certain places,' says Jennifer.

6.30, and my alarm rings. Dave the ever-watchful has already checked the generators. I waste slabs of Antarctic light-time, asleep. But darkness is brief, and losing a sleep rhythm is seductively easy. I pad quietly down the hall for my shower slot. The women's bathroom is next to the men's. Ours has posters and painted walls, a cardboard skua box for scavenging left-overs from departers, a hand's width of personal shelf space, a hook each for a towel, and a loo in a cubicle with sides that don't reach the floor. We always know who's inside. We know each other's shoes. I pray for brief privacy in the tiny shower cubicle in the opposite corner behind its flappy plastic curtain. My aim is to make my small bottle of Jo Malone shower gel last a hundred showers. The delicious smell is sanity.

Here on Station we are assailed by the sounds of the engines on which our survival depends. Antarctic scientists who work away from base live with the sounds of wind, and silence. But outsiders like me need luck. Once, in eastern Antarctica, a helicopter dropped me at a field hut, tins of food in the outer porch permanently frozen in nature's freezer, a pan of snow waiting on the floor inside for whoever arrived next to melt into water, cans of beer wrapped in a sleeping-bag on one of the bunks for the needy. The loo was a lean-to with a black plastic bag, contributions immediately frozen. Light came from the summer sky, heat from a camp cooker. Guarded and guided by a friendly field expert, I walked across a freshwater lake on metres-deep, glass-clear ice, delicate fractures thin as tissue paper, like captured coils of smoke, running deep within, and air bubbles trapped in a vertical holding pattern, graded in size from aspirin to mushroom. Strata of desiccated plants six thousand years old crumbled in the lake's banks. In the opposite direction I walked across a fjord, milky blue-white ice surface rough with captured waves as if it had frozen in an instant, and negotiated the un-anchored rocks of a hill, to look 360°: a wondering, humbling glimpse of Antarctica's great emptiness. In the distance, the waiting edge of the ice sheet, the big white as Australians say; the silence so loud it shouted inside my head.

7.10. Breakfast in Palmer's comfortable galley, with its bookshelves, cast-iron stove for real fires and strategically placed hand-basin for washing hands before serving ourselves from the counter separating the kitchen space. Paintings by artists who have spent time here hang on the walls, and sofas face south towards the sea and the sky. Breakfast is cooked, if we want cooked; there's fruit while it lasts, soya milk or long-life, fresh coffee, a drawer full of assorted teabags, a coloured drink endlessly swirling in the juicer. Bread comes generally white. Snacks fill the shelves. Or we can walk into the store and find what we want, taking it or signing for it, depending on the category.

7.30. People wander across the boardwalk to the TV lounge for fifteen minutes of stretches, included, if staff, in their hours of work. This season ten support staff are new to Antarctica. Jeff the carpenter has never until now left North America, been on a commercial flight or seen the ocean. But some of the staff have worked in US stations for years. Contracts for the Antarctic summer free the following northern hemisphere summer for travelling, on roughly a year's wages. Bob the Station Manager shovelled snow at the South Pole in spring 1993, then grabbed a chance to move on and shovel snow at Palmer the following winter. Lead summer cook Wendy, with a degree in international relations, has come south most years since her first job, bottom-of-the-rung shuttle bus driver at McMurdo in 1991. They are Old Antarctic Explorers, OAE. They've served their apprenticeship, worked their way up over the Antarctic opportunities hurdles. What matters is to get started – get south, in any capacity, connect to the internal gossip networks, find out what might be coming up, what suddenly has come up, position yourself to make the next move. Wintering is part

of the initiation. And working at the South Pole. Short-term contracts
in Antarctica can become addictive: food and accommodation laid on,
everyone healthy, no chronic conditions, no need to look after the young
or the old, potential partners checked for AIDS. Wendy: 'No unem-
ployment, no children, no crime, no income discrepancies – everyone
eats the same food, uses the same rooms. The real world is divided by
money. This is socialist, egalitarian. Living in Antarctica: it's an unreal
world.' Palmer is a small, close community, with few apparent divisions
between people. It's very popular. Wendy: 'We'll never have a piece of
land as pretty as this to live on. We are so lucky.'

8 a.m. I put on coat, gloves and boots and walk up the hill to T5.
Pinned to the back of one of the big blue doors is a small card:

This marks
64° 46' 28.01" S 64° 03' 02.45" W
62' elevation (above M. S. L.)
Facing 280° True Bearing
when door is closed more or less

T5 is important physics with no real-time scientists, an industrial site
humming with technology: solid grey floor, grey metal desks, work-
benches, stuff everywhere. Piled stuff. Old stuff. ('OK. Whose is this
little duck?' says the man from Raytheon doing an inventory, tackling
another machine.) There are instruments in glass-fronted cases. Boxes.
Loops of wire. Rolls of paper clicking out of machines, print-outs spilling from computers, sheets of paper piled in trays. Information, information. Measurements happen automatically, with results transmitted
to scientists in labs and offices on distant continents via computers and
satellites. A printer next to my right arm mimics a heliograph. Once it
was the source, linked to a piece of equipment; now its endlessly churn-
ing print-out is only significant if it stops, proving there's something
wrong with the system that has superseded it. Orion, a slender graduate
with cautious, dark eyes, is guard and guardian, as science technician.
At prearranged times he checks, moves dials, adjusts settings. Orion

has an office with a door leading to a small private deck leading to rocks and snow, and a view of the inlet. And, because we are at 62 feet above mean sea-level (MSL), occasional breath-taking panoramas of the distant mountains of the peninsula.

Space has been cleared on a desk for me, and two drawers emptied. A huge computer, presumably redundant, squats in the middle, surrounded by speakers, wires, switch-boxes and a red plastic mug with pens and a stapler, jammed. Grey power-boxes and cabling are fixed to the wall, and a thermostat set to 70°. Which it most certainly isn't. Which suits me. Orion just uses heaters when he needs. But he is a tolerant host of a stranger in his territory.

Yesterday the blue door opened wide, and Rick, storing cargo, stood in his camel-coloured overalls, big rusty beard, big lean smile, like a Norman Rockwell painting, arms wrapped around a package to add to the other packages piled temporarily on the floor. This trip I've shrunk my needs to the minimum. Racks in the cavernous store in GWR are filled with everything from light bulbs to safety pins, ski sticks to dry bags, anything removed to be accounted for and signed out on issue sheets as part of an ongoing inventory. But anything labelled 'shop stock' can just be taken. Rummaging through the shelves of old office bits and pieces, I've scavenged enough: I like their arbitrary quality. Now Rick has found me an empty carton and put a piece of varnished ply on top, and my desk area has increased by 50 per cent.

10.45. A meeting with Bob brings me back into the people-busy admin area, where my desk was last time, with its opportunity to absorb what's happening, to listen to the pulse of the Station. It works again. Bill passes, then comes in with a copy of *S-013 Extreme Biology Palmer Station Handbook*, the Birders' recipe book, the protocols, what has to be done: I've needed it. In Bob's small office heaps of papers jostle with personal things piled on the floor. Bob: 'There's nowhere else to put them. It's one of the costs of being here.' Absolutely everything essential to life and work has to be brought in from the outside world. 'Antarctic science can only happen in this context – things, and people.' Bill wants an extra hand with the giant petrel rounds and negotiates for Logistics Support Assistant Carmen. Bill: 'It's not a suntan day. It's snowing out there.' Carmen is elegant and wears fluffy angora sweaters. 'She's not bringing her knitting with her, is she?' Afterwards, Bob explains: 'If Raytheon people get out from under their day jobs and into the field, it resets their perspectives – they can connect with the science they're supporting. Understand it, so feel they are part of it. Support people need to keep that focus. As a manager, I mediate the role between support and science – balancing is a big part of what I do. You can work at McMurdo and never speak to a scientist. But here it's different.' Visiting National Science Foundation rep Dave Bresnahan, Systems Manager of Operations and Logistics for Polar Programs, is more explicit. Scientists are why the support staff are here. Scientists are the reason for their jobs.

Snow falls as I walk back to T5, past sheds and sea containers, holding piping, reels of wire, sheet metal, planks of timber, and on past the helipad still indicated on Palmer maps but now used for storing hazardous waste. Buildings and supplies are physically spaced out, in case of fire. There's no alternative down the road: anything missing is missing until the next ship. This is it, though it's easy to forget, surrounded by our carefully planned comparative abundance.

I pause to watch the ice cliffs. A Palmer habit. A blue-shadowed ice cave undermining an angle of cliff is releasing layers from its amphitheatre roof, ice skin peeling away and shattering into puffs of rainbow crystals. Everybody is waiting for the big one, the collapse of the cave itself. We loiter on the walkways, rush outside at a louder bang. Watching is addictive. But the ice performs for an unseen conductor. Decay is multilayered, and we see only the outermost edges.

A geodesic dome is being constructed across the view of the ice cliffs to hold a 9-metre dish. Currently Palmer's only link is to an old US geo-stationary military comms satellite, LES-9, sitting ten degrees above the horizon. Launched in March 1976, it's accessible for two six- to seven-hour slots a day but subject to wobbles and outages. Its 38.4 kilobits is used for absolutely everything – science data, emails, text documents, web traffic – and now digital pictures, clogging it up.

Limited connectivity does have advantages. We're here; we often can't do much about what's everywhere else. Once the new dish is installed, Palmer will be hooked totally into the outside world: unlimited internet access; live television; the ability to talk to your partner at home all day if you want; instructions from superiors delivered instantly; people connected irretrievably into offices and the decision-making process. Antarctica's isolation fundamentally breached. It's happened fast. On Australia's Antarctic stations in 1994 the latest advance – fax and telephone – had been installed, superseding the previous system of a few lines of telex using prearranged codes: YARAJ = have met with an accident WYLOP = good show keep it up. But it was prohibitively expensive. At Palmer in 1998–99 we were allocated one ten-minute call a week, over and out, via a patch into the phone system in Florida, which anyone on the peninsula could listen to. Now telephoning out is much easier. This is the last opportunity to behave in the old-fashioned way, with limited intrusion by emails, no incoming calls. These are the final moments.

A VHF/FM radio chirps on my carton, exchanges between scientists in the field and comms room crackling intermittently. Safety rules require giving destination at departure and regular reporting in, with position.

3.30. I hear Bill at Norsel Point, to Dave the comms tech: 'The wind is shifting round a bit. What's the weather like at Palmer?' 'It's OK.'

During tea in the galley the wind suddenly gets up, gusting hard. The bay runs with white-topped waves. Bill again. 'What's the wind speed at Palmer?' Pause. '30 knots gusting up to 50 knots.' Bill: 'You are describing a pretty classic weather pattern for this area. Over.' Dave: 'Suggest you head for home.' Bill: 'We are getting down off the ridge as we speak.'

T5 does have humanising touches. A few metallic Christmas decorations loop over metal poles; there's a plastic-edged mirror, the right height. Two small rectangular windows have tie-dye curtains, which I pull back to see the hard brightness of the glacier edge beyond the Back Yard ridges. Origami objects balance on the window-sill, and a label from a tin of apricot paste, bright blue, with an apricot-coloured apricot and green leaves painted in the middle. A door opens on to narrow stairs winding up to a small tower room which I never see. It must be the most private space in an endemically un-private place. I fantasise about sleeping up there in peace. But sleeping is forbidden because T5 bulges with electronics and is not connected to the Station fire alarm systems.

The wind whistles at the small windows, masking the humming technology. Fine snow works in around the blue doors and collects inside in pyramid heaps on the floor. It's very cold in this old building. I get a small sense of what it might have been like when Bill first arrived at Palmer, twenty-seven years ago.

9

An absolute wake-up call

Friday 11 January: The three-week dating season has begun. Out in the Adélie colonies young birds are arriving, wandering around, looking at vacant territory, picking a site, taking possession. Typically males three- to four-years-old, prospective breeders, they display to the females, form a pair bond, but rarely attempt to breed. The birds that lost their eggs earlier in the season are still holding on to their territories. Most pairs are still paired, going through the motions. Bill says it reminds him of people waiting for a train, some reading the papers, some looking around, or looking at their feet: no purpose. The birds are not engaged. They sit on their nests, doing pseudo-incubation shifts, exchanging nest duties but with no chicks, no eggs. They aren't feeding as much, but how they are feeding isn't known. The pairs will fall apart. There's nothing to keep them. By the end of January they will have left and the only permanent inhabitants will be the breeders.

White granules of ice tumble and meander, turning into floppy drops of rain, then wet snowflakes, and back into loose clusters of ice balls. The sky is gradations of grey; dirty grey on the horizon, white-grey above. Well-rugged-up tourists in rubber boots group attentively around Kristin the station physician, hearing how we organise our lives. Kristin smiles her instant smile beneath her hand-knitted woolly hat. More tourists trail in the wake of Brenda towards the galley for warmth, coffee and Palmer's famous brownies. Brenda's mum has FedEx-ed her some new jeans but they don't fit, so she has advertised them on the station web site, though there have been no takers so far. A maximum of twelve or thirteen tour ships anchor in Arthur Harbour every summer, and January is the busiest visitor month. Who fills the slots Palmer offers is negotiated at an annual meeting of the International Association of Antarctic Tour Operators, IAATO. Visits are resource-heavy, with requirements to lift aside furniture in the galley for coffee and brownies, 'meet and greet' guests, guide parties around the buildings, open the store, show the penguins on Torgersen. They can re-energise people on Station, and also drain. But tourism is part of the Palmer summer. Selected tour ships generally have Americans on board, giving them an opportunity to see the US Antarctic Program in action, observe how their tax dollars are being spent.

The tour ship's big zodiacs slap through the short swells to the 'open' tourist side of Torgersen Island, unloading parties onto the grey rocks. Three or four Adélies stare impassively. This morning the visitors get fifteen minutes from Bill on penguin life.

The next zodiac load waits their turn in the sleet but a lady from Florida lingers, peering up at Bill. She is determined. Bill thinks she is ninety.

'Tell me, Dr Fraser,' she says, 'what have you found? And where are you headed?'

'My life's work,' Bill says to me ruefully, 'going past at one year per second.'

Bill Fraser's entry ticket to Antarctica was a summer spent as a college senior in Alaska, night-filming brown bears with a time-lapse gun-camera from a Second World War fighter plane. His first job, training to be a park ranger, was an obvious next step on from his degree in wildlife management at Utah State. Big technology, Bill chuckles, the trucks, guns, badge and radio that go with being a law enforcement officer. Nine months in, he realised he wanted to do research in biology. David Parmelee, a genial, established ornithologist from the University of Minnesota with research experience in the Arctic, was arriving in Palmer in December 1974 to establish a biology programme. Bill got a six-week project at Palmer time-lapse-filming kelp gulls, starting in January 1975.

At Palmer, Parmelee focused on the behaviour of certain species of the area's seabirds, and assembling a database of birds with known histories by banding (ringing) selected adults and chicks. Only two activities with Palmer's Adélies were instigated, a rough annual census of breeding pairs and occasional counts of chicks. Most things that needed to be found out about Adélie penguins were thought to have been already observed or proved. Bernard Stonehouse: 'Everything, it was believed, was already known about the Adélie'. Anthropomorphism was now anathema. Birds for biologists were most emphatically not surrogate humans. Bill Sladen's initial work at Hope Bay had expanded into a detailed study, introducing modern ornithology to the Antarctic. For Sladen, Adélie *x* was a four-year-old male not yet breeding because the gonads had not yet developed. Adélie *y* doing first nest duty was a male. Adélie *z*, doing second nest duty, was a female. Insights confirmed by 'collecting' – i.e. killing and dissecting.

In October 1975 Bill returned to Palmer for his second Antarctic season. A late spring blizzard halted Captain Lennie in the supply vessel *Hero* two nautical miles off shore, and everyone had to walk in across the sea ice, Bill carrying his big green suitcase, the cook carrying a couple of 5- to 6-gallon containers of Argentinian wine and unintentionally testing the ice thickness by going through, breaking one container to general fury. Bill stayed all winter, through the claustrophobia and tension of long winter nights, the station reduced to the cook, the doctor, two scientists, two technical staff, and endless poker games.

At Palmer, Bill clashed with Parmelee. 'I had an idea of how I wanted to conduct my study of kelp gulls. He had his. I was interested in foraging ecology. What did it take for these birds to produce a clutch, raise their chicks and get through the winter? I wanted to understand their life history from the foraging perspective. Parmelee was only interested in how kelp gulls built their nests, how and when they moulted, at what time of day a kelp gull laid its egg. The goddam bird laid three eggs. I had to go out and check once every four hours over the twenty-four hours, over the period of a month. There were thirty pairs, so I had to observe up to ninety performances. Parmelee's mind was set on obtaining this information. To him it was important to know, and in this he was typical of other ornithologists.'

Lifting up gulls' bottoms to check egg production triggered Bill's interest in the growth-rate of the chicks. Chicks with parents in a good foraging territory were robust and strong. 'I began to think whether the growth rate of chicks could be used as an indicator of changes in the food supply; to wonder whether animals could be used to monitor food supply changes. I discovered that birds with only one or two eggs did not have good feeding territories, whereas those with three eggs had wonderful feeding areas next to limpets – their main food. They established prior occupation. The inferior territories belonged to young birds just establishing themselves. If an older pair died, or moved, the younger birds fought for their territories.'

Leaving Palmer in October 1976, Bill's ship was trapped in ice in the Weddell Sea. He was airlifted by stages to Buenos Aires, his first trip since the age of eleven to the city where he was born.

When did you fall for Antarctica? I ask. 'Straight away. That second trip south, aged twenty-five. I was committed.' In Buenos Aires an uncle interrogated Bill's new commitment: What are you doing this for? Why? I know everything there is to know about birds. They lay eggs. And that is it. Bill was the first person in his family to be a scientist.

The seabird work at Palmer began as a simplistic study, first-base science: an attempt to fill gaps in knowledge about Antarctic seabirds. Bill: 'Parmelee developed his science in the vacuum of the lack of the budding new science of ecology, a revolution in ideas. I was lucky. I was studying at a time of great ferment. I slipped into a decade when so much theory was being developed. But – the old data could be placed in the new context.' Parmelee's insistence that students went out and documented the detail meant achieving figures, data. His forcing students to perform repetitive tasks instilled a daily basis in the field. 'Being in the field is essential given the amount to be described, to absorb, and observe. In the field you are presented with an infinite number of events. The information is out there, floating around. It's a cloud at your doorstep. The trick is to know something well enough to pick out the events, the patterns, that lead to another brick on your pile. Parmelee's training has persisted to this day. It is the basis for everything we do.' It went into creating Bill's own protocols. It ended up forming the backbone of his methodology, the long-term databases he has built up.

Tracking Bill's work means, for me, unpacking strands. Information can come unsequentially. Some gaps I don't fill until long after, some I never try. Much of what follows I learned during my first summer at Palmer. My desk was ten steps from Bill in the chief scientist's office. The photocopier clacked just outside, coffee filtered in the galley, a plate of freshly baked cookies was always set out on the counter. Ron, the Station Manager, wore shorts and a T-shirt, and a door newly knocked through at the end of the corridor stayed open to Antarctica. The companionable snorts and snuffles of three young elephant seals hauled out on the rocks below permeated gently through our spaces.

Part way through my Palmer time I joined the British navy's ice patrol vessel HMS *Endurance*, to experience their work along the Antarctic Peninsula. An American penguin researcher working on board consigned me to a camp I didn't know existed, because I'd 'come from Palmer'. Puzzled, on my return I asked Bill to explain. And Bill began unravelling the biography of his ideas, unpicking his hypotheses. Like most busy scientists, he hadn't had time to stop, to describe the process.

Bill: 'Climate warming didn't enter anyone's reckoning when I was training. It was not something you would observe in the fragment of time your life occupied. It was a real process but it was documented in glaciology, in geologic time.'

But in 1977 a rigorous study of global warming undertaken by the US National Academy of Sciences unequivocally linked rising levels of atmospheric carbon dioxide with climate change, concluding that if emissions continued to rise at their current rate temperatures could rise, with serious consequences. In 1979 the Academy asked a panel of experts to check the results from early climate models that looked at the effects of increasing carbon dioxide in the atmosphere. The panel accepted the models' calculations. A wait-and-see policy could mean, they warned, waiting until it was too late. Pressure to understand the Earth as a single integrated system, a single entity – to study the planet at a global scale and in a multidisciplinary manner – increased.

Traditionally, the Office of Polar Programs at the National Science Foundation (NSF) in Washington DC funded research projects lasting two to three years. Scientists defined a hypothesis and proposed ways to try and answer it, requesting funding to implement the research needed. It became apparent that questions relating to developmental biology, or climate, required a longer time scale and a different kind of funding. NSF set up a Long Term Ecological Research (LTER) division, with mainly terrestrial sites. Phytoplankton expert Maria Vernet: 'Once in a while we do a big jump. It's a rare event. The birth of the LTER represents one of those shifts.'

Extreme regions were central in the search for the mechanisms of Earth's changing climate. One line of research was to look at the role of Antarctic sea ice. Bill: 'The huge differences between winter and summer coverage, for example, must have important ramifications to sea ice as a habitat. I was a beneficiary of this attitude, in the right place at the right time. I was just coming in as a player, I had winter experience in Antarctica plus work with Adélies at Palmer.' Major illness had taken Bill out for four years, 1978 to 1982. The following five years had been mainly spent working at sea for seabird ecologist David Ainley, visiting Palmer briefly during research cruises. Now, during the 1987–88 season, Bill took on much of Parmelee's work at Palmer.

In 1988 NSF chartered the M/V *Polar Duke* to undertake one of the first Antarctic winter research cruises, surveying 581 square kilometres of the Scotia and Weddell Seas, strip-censusing through pack ice and open water while researchers recorded abundance and distribution of seabirds and marine mammals. Bill was on board. He laughs – 'an announcement came over the ship's systems, "will all scientists, and" – pause – "biologists, come to the bridge".'

In the sea ice the ship's spotlights picked out thousands of Adélie penguins standing and lying on the floes in the darkness. But chinstrap penguins were observed swimming by the thousand in open water. Adélies and chinstraps both eat mainly krill in their summer diets. Their general appearance and size are similar. They have broad ecological similarities. Yet here they were, occupying very different habitats in winter. Bill: 'The sight was an absolute wake-up call. A major turning-point in my thinking.'

In Antarctica penguins were considered to be 'indicator species'. The food chain involved was thought to be remarkably simple. Large animals at the top, such as whales and penguins, ate krill, the small shrimp-like *Euphasia superba* which fed on phytoplankton, the grass of the sea. The hypothesis was that, as krill-eaters, penguins could reveal if too many tons of krill were being hauled out of the southern seas by the proliferating Eastern European fisheries: declining numbers of penguins would indicate too much fishing. Research was being carried out on penguin numbers along the Antarctic Peninsula, focusing on the five inner-island Adélie study sites at Palmer and a mix of sites further north at King George Island in the South Shetland Islands, where Adélies, gentoos and chinstraps nested. Concerns about managing resources drove the research, as well as concerns about changing the ecosystem.

But there was also the whale population reduction hypothesis: not a krill deficit but an abundance. So many krill-eating baleen whales had been hunted and killed since the 1920s that there must, it was argued, be a 'krill surplus' now that hunting had declined.

The problem was the figures. At King George Island the data of American researchers Wayne and Susan Trivelpiece showed the numbers of chinstrap penguins increasing, while the numbers of Adélies see-sawed. The chinstraps' range was expanding south down the western side of the Antarctic Peninsula. Abundant supplies of krill, it was argued, would account for the increase. But Adélies were also krill-eaters. And at the Palmer study sites Adélies were in decline. Their numbers had been steadily decreasing from 15,202 breeding pairs in

1975, when the data set began. Bill: 'Intuitively the numbers did not fit together. Hypothesis. Data. The two didn't mesh. Classic science. I had pieces of a puzzle floating around in my head. I had a bunch of numbers lined up into a graph and I didn't know what they meant.'

Why were the Adélies at Palmer in decline? What were the factors affecting penguin survival?

Bill says his next insight came immediately after the 1988 *Polar Duke* winter cruise, at a symposium on Antarctic biology organised in Tasmania by the Scientific Committee on Antarctic Research (SCAR). Listening to data on population increases and decreases among various species provided a 'quantum leap' in Bill's ability to understand. What if the winter habitat of Adélies and chinstraps was significant? The winter cruise had shown that Adélies were obligate inhabitants of sea ice. 'Adélie numbers were decreasing, chinstraps increasing: were we seeing a change in the relative availability of the two species' habitat? That was the catalyst thought. If so, what was the mechanism for the change?'

Overwinter survival is crucial to penguins. Dead penguins can't come back to breed. Underweight penguins with insufficient blubber insulation cannot sustain the necessary brooding fasts. Overwinter survival, to use scientist-speak, can play a key role in driving long-term populations.

Central to Bill's argument was the quantity and extent of sea ice. Had it changed? Was it reducing?

10

Light bulb moments

At Palmer, and afterwards back in Cambridge, England, I searched for the tenuous, sometimes chance, actions that helped underpin what is currently known. Why was one kind of science in Antarctica done over another and what influenced success? How was the doing of science influenced by the reality of people's lives in these isolated, remote, narrow communities, set in the midst of existential grandeur? How random were the value and quality of this sole Antarctic export, science? The human history of Antarctica is brief, and fitfully recorded. Archive research and interviews gave me connections between the observations by Base N men of the local biology, and Parmelee and the start of Palmer research. But a key measurement for climate is surface temperature. How did essential data sets for surface temperature in the Antarctic Peninsula region get established? Climate change studies require finding out how something was, before the way it now is. What was involved in collecting data of sufficient quality to be useful as a climate record?

All the early scientific and national expeditions to Antarctica achieved some meteorology, along with biology and geology. Standard scientific activities, they were accessible and inclusive: anyone could contribute, results were guaranteed. Geology's advantages were undisputed: locating mineral resources was a clear priority. Biology was everywhere for the taking. Meteorology was an established activity, procedures in place, observing and recording carried out on long sea voyages by deck officers who drafted in anyone with met training, and on land by a meteorologist or whoever was available. An internationally pursued activity, meteorology was part of the revolution of measuring that began in the seventeenth century, involving precisely calibrated, expensive instruments with components of glass needing careful transport and handling. Instruments of status, proof of European scientific values. Figures noted down in log-books, records kept.

Tucked into a comfortable chair in the British Antarctic Survey (BAS) canteen in Cambridge, back to the window, seniority clear, the fingertips of his surprisingly small hands lightly touching, Brian Gardiner explained to me how, instead of doing physics as a newly arrived physics graduate in Antarctica in 1967, he found himself one of a group taking met obs: a set series of measurements and observations at three-hourly intervals through the twenty-four hours of every day, of every year. Home was a wooden hut on a narrow raft of black rock surrounded by abrupt mountains and tight steep-sided fjords: Base F, 51 kilometres south of old Base N at 65° 15' S, 64° 16' W, established as an atmospheric station in January 1947. Brian is one of three British Antarctic Survey authors of the 1985 paper revealing the dangerous thinning of the ozone layer – research arising from long-term instrument readings, taken regularly over the decades mainly at Halley Station on an ice shelf by the Weddell Sea, but also at Base F (later called Faraday and now transmuted into the Ukrainian station Vernadsky).

There were Adélies at Base F, but no one studied them. Brian: Sometimes, out where the Station huskies were tethered, a lone Adélie would come wandering within range. I used to wonder, why do penguins do it? They just walk straight into the middle, the dogs wait, then wham. One penguin torn to bits. So I moved fast, kicked the bird onto its back and, as it turned over, picked it up, holding the flippers firm against the body like a package, and threw it off the cliff into the sea. One penguin saved.

Met obs meant going out to the met screen twenty minutes ahead to put the wick of the wet bulb thermometer in a little bottle of distilled water, unclipping the white-painted Stevenson screen exactly at the hour, dry and wet bulbs read and the lid closed all in thirty seconds, breath held to avoid affecting the result. Recording temperatures in a notebook. Repeat the same procedure on the two thermometers constantly bathed in fresh air sucked up open-ended stainless-steel tubes in the second white box, the Assmann. The night met man took the cloud searchlight with its 500-watt bulb, pointing it up vertically to read the clouds, memorising the estimates of cloud cover and height. Back inside, outdoor clothes hung up, then into the little narrow Synop room to record all obs in the Meteorological Register, kept open and ready on a bench, landscape format, stiff cardboard covers, one spread per day, laid out with almost legal precision. Page after page of neatly filled-in observations, written in ink, signed off by whoever had been the met man, no place for error, no opportunity for corrections: 'This is what I observed, at this time, in this place, to the best of my abilities.' Columns for fifty possible observations, highly standardised, everything to be coded, code books on the shelf for reference. Temperature. Pressure. Precipitation. Cloud cover. Calculating humidity. Dew point. More obs recorded from instruments on the wall – anemometer, barometer, wind direction dial on the barograph. All codes, noted again in the Ops Book then taken next door into the Ops Room for the radio operator to tap out in morse every six hours to Stanley, in the Falklands, to go into global weather forecasting.

Night met lasted a week at a time, obs at 9 p.m., midnight and 3 a.m. All observations were visual and manual, all calculations made with hand-turned calculators. After recording the last obs 'I'd check outside to note present weather. Check around the base, because night met is night watchman. It's not a bad rota. Bed when the day met man comes on to wet the wick at twenty to six, before he takes the 6 a.m. obs.' Except of course – Brian gently smiled – the met man was living at 'kitchen time', local time, which was three hours behind scientific time, zulu time, GMT. So his three night-time obs were recorded in the Meteorological Register as occurring at midnight, 3 a.m. and 6 a.m. zulu.

In 1967 Brian and his five fellow researchers were starting the twenty-first unbroken year of meteorological observations at Base F. Brian: 'To be of real value measurements need to be a long-term sequence.

Understanding comes from continuous measurements, repetitive, consistent, measuring the same things in the same way at the same place, with no gaps. The only thing better than ten years to a meteorologist is twenty years. But if you want to establish something – to look for a trend, a pattern, changes – you need thirty years.'

In 1988 John King took over as head of meteorology at the British Antarctic Survey, inheriting a project from his predecessor, David Limbert. 'There appears to be something of interest in the Antarctic Peninsula region,' Limbert told King, 'some interesting variability.' Limbert, with Phil Jones from the University of East Anglia, had begun pushing all the air temperature figures for any Antarctic stations they could get their hands on into a computer, looking at a climate change scenario. The search for temperature changes was part of basic climate-related research funded by the Carbon Dioxide Information Analysis Centre in Oak Ridge, Tennessee. Computers were no longer fridge-sized boxes in air-conditioned rooms. Computers had become smaller, accessible, affordable and increasingly powerful, allowing large quantities of figures to be processed by researchers.

Over coffee at BAS, John took me through the background to the climate warming figures for the Antarctic Peninsula. 'There's no story if people write down the numbers and file them away. In order to make progress you need to dust off the boxes of data. Subject them to quality control. Check if the readings are believable. Look at the met work of similar stations in the region. Exclude dodgy data.'

But some of the climate records from some Antarctic stations were, in John's careful phrases, 'patchy, of questionable quality'. After the Antarctic Treaty came into force in 1961, with doing science the justification for being in Antarctica, meteorological measurements became for many participating nations an ideal 'bit of science' for credibility. 'Taking met readings was considered a good thing to do partly because you could. And partly because it's what you could do if you wanted to be seen to be doing science. Even if you had no idea of what you were doing. Collecting simple meteorological data was the most straightforward of scientific recording. It didn't need scientific experience.'

But it did need relatively rigorous procedures. Swaths of data had to be excluded: a mix of gaps in the observations, insufficient staff, dropped stitches, lack of motivation or commitment. Some countries wouldn't hand over the figures, despite the agreed terms of the Antarctic Treaty: records, they claimed, belonged to the armed forces and were deemed to have military uses. A long enough time sequence was necessary for any conclusions of value, because of anomalies. John: 'Things can fluctuate randomly. Variables can mask a trend. A trend can just have been chance.' The figures were subjected to a series of tests. BAS meteorologist Tom Lachlan-Cope: 'The variables in the Faraday figures didn't start to cross the border into being statistically significant until the 1980s.'

The Western Antarctic Peninsula turned out to be warming rapidly: a summer warming of 2.5 °C over the previous fifty years, but a startling increase of 4.5 °C in the mean winter temperature. The region was experiencing strong long-term warming trends, and a much greater degree of variability, than anywhere else in Antarctica. That made it one of the fastest-warming regions on Earth. The surface air temperature records taken without a break, every three hours every twenty-four hours of every year at old Base F, which had become Faraday Station, were a vital unbroken sequence. They were a crucial resource.

John: 'A lot of people had noticed the extreme variability of the Antarctic Peninsula climate on the western coast.' Papers began to be published from 1987.

Bill described to me how, after the 1988 winter cruise and the biology symposium in Tasmania, he moved his thinking on a step, using the newly available temperature data. Applying them, jumping scales, looking at populations from local to regional to global scales, he developed a hypothesis. 'Was it possible that we were seeing the effects of climate warming in the decline in Adélie numbers and the increase in chinstrap numbers at Palmer and King George Island?'

The mechanism Bill was proposing was a reduction in the quantity and extent of sea ice in the region, with climate warming. But was that happening? It was vital to find out. Bill: 'I didn't know if sea ice had decreased. I thought it could have.'

Antarctica and the Southern Hemisphere

0	Statute Miles	3000

0	kilometres	5000

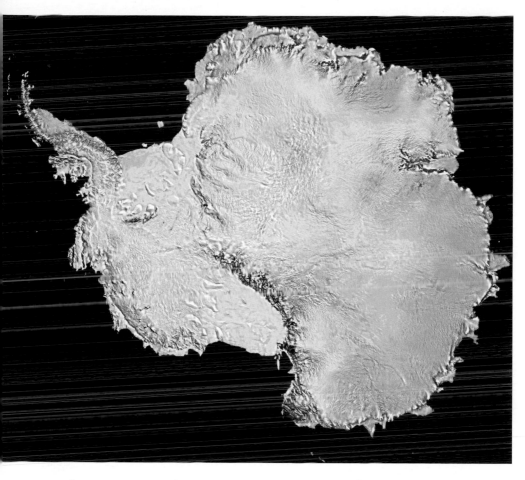

1. *Like a giant stingray the Antarctic continent heads across the Southern Ocean.
Two enormous icebergs detached from the Ronne Ice Shelf float in the Weddell
Sea; several from the Ross Ice Shelf can be seen just off Ross Island in the Ross
Sea. This cloud-free mosaic of Antarctica was developed over several months from
260 separate digital images acquired by the MODIS sensor on board the Terra
satellite. The map opposite shows the Antarctic continent in the Southern Ocean, in
relation to southern hemisphere continents and islands.*

2. *Palmer Station, Anvers Island, during the ferocious summer, with the Back Yard and glacier behind.*

3. *Looking across the Marr Ice Piedmont on Anvers Island to Mount William, 1,515 metres high.*

4. Leader of the sea bird team at Palmer Dr Bill Fraser selecting an Adélie on Humble Island. Pinky brown guano shows that the penguins are eating krill.

5. A Humble Adélie fitted with one of the precious five functioning satellite transmitters in January 2002.

6. Ferocious summer Adélie chicks crèching. The smallest has hatched too late and won't survive.

7. A chick in supplicant posture, about to receive food.

8. *Gentoo penguins are arriving at the Palmer islands in increasing numbers.*

9. *Chinstrap penguins have been observed nesting in the Palmer area since the mid-1970s.*

10. *Elephant seals swim south to the islands around Palmer every summer to haul out and moult.*

11. *Young male fur seals arriving on Laggard Island, note the brash ice in the sea.*

*12. Brown skua, wings full stretch, menacing an Adélie chick, with an adult.
Bottom photograph, skua larder – skuas feed on dead Adélie chicks when they
need. The band on the left leg means that this bird was banded as an adult.*

13. *A male Adélie calling with head raised and flippers extended, announcing that he has a nest and no partner.*

14. *Adélie colony on Humble Island with elephant seals, Palmer Station and the peninsula mountains behind, distances compressed by telephoto lens.*

Resolving the question of decreasing sea ice was exceptionally difficult. No adequate database existed. There were sporadic reports of ice extent observed from ships, especially whalers. But the evolution of satellites was starting to provide a second technological enabler, to add to the revolution in computing power. Satellite images beginning in the 1960s did not distinguish the mass/detail of sea ice, so no measurement was possible. But in 1979 microwave sensors attached to satellites started delivering, and data sets of sea ice extent began to build.

John King: Sea ice decline is complicated. A lot of inter-annual variability is characteristic of the Western Antarctic Peninsula. With any climate variability, if you are looking for long-term change, the signal is buried beneath year-to-year variability. The sea ice duration record has only recently become long enough to see if there are any systematic changes.

Tracking his hypothesis, Bill applied for information to NASA Goddard Space Flight Center, outside Washington DC. 'They said categorically there was no change in the sea ice. There were too many years with too much sea ice, too many with too little. There was "too much variability in the system" to produce a trend line.'

Here, said Bill, was his light bulb moment. This was a question of frequencies of years: variability could be part of the formula. He focused on whether the variability had changed, looking for some mechanism for change. He correlated the temperature record for the region, as detailed by the British, with variability, and found that in the 1950s four out of five years were cold. Now only one or two were cold.

The paper setting out the thinking used the phrase 'environmental warming' in the title. Bill said *Science* and *Nature* wouldn't review the paper. The day that a seabird biologist can tell us something about climate warming is the day we are all in trouble, responded one editor. Bill: 'That's the gist. I can look up the exact words. But it's pretty close. It was crushing. Current belief was that geologists and glaciologists would inform us about such long-term issues as climate warming. I didn't fit the classic role.' The paper, co-authored with US Adélie specialist David Ainley, working in the Ross Sea region, and with Wayne and Susan Trivelpiece, whose data on the increase in chinstrap penguins at King George Island had recently been published, was rewritten and published in 1992 in *Polar Biology*.

The paper proposed a mechanism for linking changes in penguin population numbers to environmental warming. It hypothesised that environmental warming was changing the availability of critical wintering habitats for both chinstraps and Adélies. The patterns of recruitment and population growth were altering. The answer to the increasing numbers of chinstraps did not lie in the idea of a 'krill surplus'. Chinstrap penguins were increasing because of a gradual decrease in the frequency of cold years, with their extensive sea ice cover. Given the complexity of understanding the relationships between changes in biological populations to ecological perturbations, the paper concluded that the management of Southern Ocean resources should pay close attention to environmental data.

Bill: The paper was considered extremely controversial, causing deep rifts in the small world of penguin researchers. Battle was joined over priority, over the funding of programmes. 'Politics intervened. The paper struck at those whose whole careers had been invested in the single thesis that the marine ecological system was the only explanation for penguin populations. I have been attacked ever since. I had put forward a hypothesis. It was disagreed with. There were attempts to derail it, but also to claim it. A valid scientific debate has gone to a personal roast.'

At Palmer Bill had taken over Parmelee's work completely in 1990, inheriting the data. That year, NSF set up a polar program within their Long Term Ecolgical Research division to study how sea ice impacted the functioning of the entire biosystem. The decision was made to base the Polar LTER at Palmer. Maria Vernet, Principal Investigator of the phytoplankton research at Palmer, put the evolution of Bill's science with admirable simplicity: Bill started looking at birds. But birds are long-lived. So in a certain way he already had long-term research. When the LTER program began at Palmer as long-term sampling to answer long-term questions, Bill's work was adroitly folded in and

Adélie penguins, as key avian predators, became the lead animal of the Palmer LTER. Through participation in the LTER his birds are part of the bigger system.

After the large-scale patterns of the climate warming paper published in 1992, Bill said he focused his research down, tight, onto the local landscape, the local habitat at Palmer. He wanted to understand how landscape can influence the various parameters of biological populations. At Palmer every Adélie colony is isolated on an island, only accessible by boat. To Bill their surfaces are his work-bench. 'Palmer gives me pure landscape. As pure as I could ask for. I have seventy-two penguin colonies, each occupying a different piece of landscape. Each in its detail, its specific qualities, is in effect an experiment.'

By 1992 the number of breeding pairs of Adélie penguins at the five inner-island study sites had dropped to 11,974, from the 15,202 in 1975. But the rate of decline differed between islands. Why? Were humans impacting on the penguins? Human disturbance is a controversial issue. Questions of tourism are involved. There are environmental, economic, conservation and political overtones.

Anecdotal evidence for past human impact on Antarctic penguins is clear. Penguins were harvested as a resource for oil, human food and dog meat. Their eggs were eaten. Adélies, gentoos and chinstraps need bare ground for their nests, ice-free, with glacial gravel and access to their food in the sea – the kind of landscape also targetted by explorers and scientists for base camps and huts. Humans and wildlife both tend to converge on that small fraction of Antarctic coast that is free of ice. Buildings, rubbish dumps, dog spans, paths and roads, an aircraft runway, have all been put on top of penguin nests. Helicopters have over-flown them. Some scientists persevered with invasive investigations. Others required death to establish conclusions. And people liked looking at penguins, which could mean getting too close. Curiosity, enthusiasm, ignorance, callousness, research, greed, profits, hunger and carelessness all took their toll. But by the 1970s, post various provisions of the Antarctic Treaty, human disturbance studies had been narrowed down to two sources: tourism and research.

At Palmer the availability of Adélie breeding habitat has never been altered by human interference. Buildings have not been put on nest sites. But after 1975 human activities did increase significantly, and from 1978 the NSF Polar Programs section initiated tourist management policies that reduced access to penguin colonies, limiting tourism unofficially at Palmer. From 1986 no more than 1,300 tourists were permitted to visit each year, and from 1990 they were permitted access to only the 'open' half of Torgersen Island, with the same restrictions on Station personnel during the nesting season, 1 October–1 March.

Here was a natural laboratory, providing Bill and his team with a natural experiment. The effects of human activity on visited colonies (experimental) and non-visited colonies (control) could be monitored. Actual visits could be measured, to reveal potential impacts from both sources, tourism and research, at specific islands.

The figures did not correlate with recorded penguin losses. Litchfield Island, a Specially Protected Area since 1975, had minimal human disturbance – no tourism, negligible research activities – yet it was experiencing the most severe decline. The island also had the largest number of extinct colonies in the area. Tourism and research-related activities continued to increase on Torgersen, yet the decline in penguin numbers was much less severe than on Litchfield. Between 1975 and 1992 there was a 19 per cent loss of Adélie penguins on Torgersen, and a 43 per cent loss on Litchfield.

Bill: 'If you took the big themes put forward for the decline in Adélie penguin numbers – fisheries, or the impact of tourism – you could look at the data from different Palmer islands and realise that those big themes would not be relevant at, for example, Litchfield.'

Bill argued, instead, that complex environmental factors could force changes in Antarctic penguin populations. Litchfield Island has steep hills. The entire rookery, active as well as extinct colonies, is located on the south-west side of a series of high rock outcrops. At Palmer the predominant wind direction during storms is from the north-east. Strong winds scour snow from surfaces, dumping it downwind. Penguin colonies in the lee of high land are vulnerable to heavy snow accumulating during the vital nesting period. Bill noted that 86 per cent of extinct colonies on all islands, when compared with currently active colonies, were located on the south side of topographical features: downwind. Torgersen Island, on the other hand, has all of its currently active colonies directly exposed to the scouring northerly winds.

With increased warming on the Antarctic Peninsula the mean annual snowfall in the Palmer area was rising. Storm systems came from the north, making breeding sites with a south-facing aspect, where snow accumulated, very vulnerable. Bill identified a long-term change in the pattern of snow accumulation as a key agent in the declining populations, citing evidence from abandoned and extinct colonies as support for this hypothesis.

But a second factor for Litchfield's declining Adélie numbers was kicking in. Brown skuas are powerful predators at Palmer, eating exclusively penguin eggs and chicks during the breeding season. More than half the area's brown skua pairs are on Litchfield, and predation pressure is high. Predation from brown skuas aided the 'final step' in the extinction process. Data from Litchfield showed that once a colony fell below about 25–30 pairs, the skuas took every egg or chick. If two such seasons followed each other, the colony went extinct. Long-term changes in the pattern of snow accumulation, combined with the effects of brown skua predation, seemed to be key agents forcing the Adélie population decline on Litchfield Island. Bill: 'Every island is a precise landscape. Litchfield is a protected site, so human impact is minimal. The decline in penguin numbers has been the effect of its landscape.'

The conclusions were clear. Potentially adverse effects of tourism and research might be negligible when compared to the effects imposed by long-term changes in other environmental variables.

Burned by continuing attacks on his 1992 environmental warming paper, Bill says he'd learned cunning. That first paper had used up so much energy. He hid his thinking under the phrase 'human disturbance' in the title of an article on the effect of landscape on the breeding success of Adélies at Palmer, co-authored with Donna Patterson and published in 1997. A paper published the year before, co-authored with Wayne

Trivelpiece, was an attempt to synthesise the ideas in both papers.

The 1992 paper had looked at what might be driving the system. On the big scale, Bill had hypothesised changes in the extent of sea ice. On the small scale, he had hypothesised changes in the summer breeding habitat of the Adélies. For the precise focus of the local Palmer islands, that meant the effect of climate warming on a critical component of the Adélies' breeding cycle, their summer nesting sites and on a critical component of their winter survival – the presence of sea ice. Bill: 'The 1992 paper was not the first to say that climate change was a possible cause. A lot of work had been published proposing that climate change may be having an effect on penguins. Our paper was the first to propose the mechanisms: the effect on summer habitat, and the effect on winter habitat – that is, changes in the relative availability of habitat in winter. It's a pivotal difference.'

Then, suddenly steely: 'One thing I'm rabid about. I'm protecting my ideas. Ideas are the currency by which scientists measure you and by which you are remembered when you are gone. The issue of interaction with people is to get straight where ideas originated. How ideas are tested, accepted, rejected – it's irrelevant. Others might think differently. The root of conflict has been the ownership of ideas.'

Andy Clarke, independent-minded British Antarctic Survey ecologist: 'Science in Antarctica is different from other places. People will tell you how science works, how it is organised. But they don't tell you how they fight. There are clashes of personalities. Styles. Ways of doing things.

'In Antarctica science and the place itself interact. Antarctica takes up more of you.'

11

Collecting birds

Saturday 12 to Monday 14 January

A snow petrel lies on the bench. The bird I so admire in the pack ice, moving so freely. We're in the Birders' lab, the end space, sliding doors leading out to the deck and the tent. I hold it on my palm. No weight, life gone. Attached to one delicate leg, tied on by string, is a brown label with details given in the way they have always been given, in pencil, or indelible ink. *Pasadroma antarctica*. Place found. Description. It flew into the superstructure of the *Gould*, and died, and a delighted scientist preserved it for the American Museum of Natural History. A snow petrel has rarity value. One day the scientist will come back and pick it up.

Bill is surprised I don't know that this is a skin. All ornithologists know what a skin is. I'd imagined a flattened shape like a bear rug – backbone centred, legs and arms stretched out in symmetry – or rolled up, like a parchment. Instead, here is this softly rounded package, a slightly elongated, narrow, column-shaped bird, lying on its back, legs crossed exactly so, like an Egyptian mummy lying straight in its coffin, or a Dutch painting, flowers, fruit and a bird on the darkly reflective surface of a table.

How do skins happen? Bill patiently explains. A slit is made along the body and the skin peels away from the insides like a glove coming off. All that's left are the bones of the skull, and the leg bones, broken at the joint. It's preserved by borax and doesn't decay.

Older ornithologists have collections of skins, drawers filled with birds in all stages – moult, growth, age and colour, sex and type, all for identification. Museums have massive quantities; some arranged in tall glass cases, for edification, with painted landscapes and props of moss, twig or rock. Symbols of how things were done, but also of how things were discovered. The actual object, the physical presence that is still wanted, still desired. The hard copy. And, of course, now I think about it, Darwin's birds collected on the Galapagos Islands, contributions to *The Origin of the Species*.

'To collect' is shorthand for kill. When did Bill stop? 'In 1993. We needed to look at the stomach contents of flying birds in winter time, and that meant killing to obtain samples. We took the information we wanted then shipped them to a museum.' Why stop? Bill: 'Evolving consciousness. The problem is I've gotten to know the birds here so well. They have an individuality. Daily stresses. They have lives – at whatever level an Adélie or a giant petrel conducts its life. Plus we now know that all these birds we work with are very long-lived. They have traditions. Mates that don't see each other for months. That is the issue now.'

Did Parmelee change? Bill: 'He spent his lifetime here. He got to know these birds as well as anyone and was still willing to collect them. It's a personal decision. People still collect. We all have similar training. Museums still want skins. There's still a value in looking at study collections.' Skins offer insights into anatomy, DNA, behaviour.

I think about study collections as outgrowths of seventeenth-century cabinets of curiosities, swelled by links to the amateur tradition, because anyone with a gun, an eye and an interest could sort and keep; swelled by links to the collector – that condition we recognise in ourselves to possess things that are limited, rare or beautiful, to give value – monetary, psychological, cultural, aesthetic – to the owner. The sheathbills under the trash incinerator with their one small, shiny key. Here in Antarctica scientists essentially still collected: krill, icefish, foraminifera, the 'mash-n-grab' guys, finders and grinders who dive under instruction for biochemists, collecting from the sea floor, or just above, in the hope of discovering useful genes and gene products. The portcullis only seems to have dropped selectively.

We go up the narrow stairs to the small crowded comms room to study the weather data. Weather rules in Antarctica, but Palmer weather has an extra dimension. It can be claustrophobically local. Blue sealed drums with emergency tents, stoves and food are anchored securely on selected islands. Before we can even sit in a zodiac we have to pass a written test, Boating 1, and be instructed how to put up the tents, use the stoves. The sea whips up quickly, in sharp short swells. Safety regulations specify (generally) no boating in wind over 35 knots.

Print-outs of satellite pictures from the TerraScan antenna up the track show the fronts building and departing over the Antarctic Peninsula. But Bill defers to the analogue anemometer, with its sensor on the antenna tower on the roof above us linked to a pen tracing wind speed and direction in real time on a sheet of paper. When the wind begins switching direction erratically, the weather will change. Fast. He shows me where it did so on Thursday afternoon when he had to return from Norsel Point – a kind of tight horizontal agitated scribbling back and forth across the centre line, followed by a steady tracing on the right. 'Keep watching the lines on the tracing. People don't believe me, but I know.' Me: 'You're a resident'. Bill: 'It's the same as being a farmer. They know the local weather.'

Early this morning we woke to see a two-masted, steel-hulled 72-foot schooner unexpectedly moored in the inlet, dimensions perfectly fitting the proportions.

Hospitality in isolated Antarctica isn't straightforward. Potential hosts have rules. Unscheduled visitors? That depends. Palmer is on a yacht route and doesn't provide facilities for stray tourists or adventurers. The same reasoning keeps most explorers who attain the south pole outside America's Amundsen–Scott station. Dave Bresnahan: We are not running a rest and recuperation station. But a member of a British Army Antarctic Expedition on the *John Laing* has broken her thumb mountaineering. It's a medical emergency. Traditionally everyone helps in Antarctica, and permission has been given for an X-ray. Skipper, Doctor and expedition leader come ashore but, not being on the list of visitor types who can be offered lunch, they are given coffee and cakes, the short version of the Station visit, and parked in the bar.

The unlucky 23-year-old's thumb turns out to be badly broken. It has to be reset, and suddenly the day takes a new shape. We can accept the invitation to visit the schooner, an old-fashioned youth training vessel, and glimpse the reality of cramped lives, with clothes washed in a bucket failing to dry. The expedition has taken ten years organising and fund-raising. The skipper's officer wife was meant to be on board but was posted to the war in Afghanistan. Consciousness of that conflict doesn't really penetrate Palmer. Our news is a short daily printout of an American news-sheet with (largely) general 'human interest' pieces.

Because it's Saturday there's an hour-and-a-half of House Mouse to get through first, cleaning tasks allocated by Brenda – shower walls, lab floors, every inch of the bar, sweat off the gym equipment, kitchen equipment taken apart. Our berths and bathrooms are meant to be cleaned mid-week as well, and every evening four of us sign up for an hour's 'gash', naval slang for rubbish, transmuted on some Antarctic bases to mean domestic chores. Here at Palmer that means cleaning kitchen, galley and pantry, a mammoth washing up, leaving everything ready for the next day's early morning cook. But House Mouse is the big one. There's no one else to magically clean up. This isn't a hotel, it's home. I think of the surprise on the face of a new inhabitant walking into the lounge the morning after the night before to find it littered with popcorn, down comforters, dirty glasses, videos out of their covers, blinds closed. With no one to do anything about it. Except, as the realisation dawned, us – which included him.

Permission is granted for the sixteen expeditioners to come ashore for their first showers for a month (Dave Bresnahan: It's because they didn't ask), and for drinks and 'cross-town' pizzas in the lounge. The evening lifts into a good party. And at the end the bearded Major pipes everyone along the boardwalk, across the rough ground, measured tread, eyes ahead. Then he stands playing, silhouetted in his kilt against the water's edge, framed by the quiet inlet of sea, low rock promontory, distant mountains. The sky is an indefinable, high-latitude light, pale green with delicate indigo. Water laps gently. It could be the highlands of Scotland, the old world of the north, touching the barely known south.

Ships' captains gave Scottish names to island groups off the Antarctic Peninsula: the South Orkneys, the South Shetlands. The music is right for the space. Lonely. For those with the associations to take hold, the heart-sad music of farewell.

Sunday's tour ship has an inflatable cartoon character strapped to its bow and eighty American East Coast high school students are on shore. Bill and I decamp to the tent. Warning lights from machinery backing on the road outside flash through the fabric, and passing feet thud across the decking. Bill: 'We are not entirely unhappy with the tent as an allocation. It confers a kind of moral high ground, helpful in negotiations. And if you look in at 10.30 a.m. we are all computing, but there's constant banter and interplay A lot of fun. Private fun but out of sight. We can discuss people. Have a complaining session.'

Then, flatly: We are in the process of documenting the failure of Adélies to reproduce. There are no final figures until February – but 70 per cent of the birds that arrived have failed. The extreme conditions have created a bunch of problems. We don't know the solutions.

The harsh early months of spring and summer underlie all the current seabird work, and the work ahead. Hard decisions need to be made. Every January for example, intensive diet studies are carried out, involving capturing Adélies and lavaging their stomachs. Bill: 'The ethical issue is: diet samples are traditionally only from breeders. But how do we take diet samples given there are so few breeders? How much do we want to disturb these birds beyond the disturbance they are already having? I have a choice. I can ignore the ethics entirely and go through with the same procedures and do the same numbers. But how much more stress can these animals handle? It's not easy. Overall, we think there is food in the environment in reasonable quantities for the Adélies. But food is not the limiting factor. The problems are the sheer impact of flooding, snow, predation, ticks – although ticks are no longer an issue. Plus, when the non-breeding birds leave, the chicks will be targeted by skuas – although the pressure may come off at some point if the skuas themselves have fewer chicks.'

Significantly, the modest number of established chinstrap penguins are holding their own, but the gentoos, increasing every year, are doing exceptionally well. Chinstraps and gentoos, quasi-temperate species nesting further to the north, were very visible in the Base N years, 1955–57, but the first chinstrap colonies on the Palmer islands weren't censused until 1977. No gentoos are nesting on the inner islands but Bill says a huge number of gentoos have invaded the snow-free areas. 'They are walking over the colonies, which are ideal breeding sites. We have never seen so many here. It's a unique observation, random, but not without context because of the experience we have of this area. We don't know what could happen. If these birds pair, they will be back. Their presence is another indication that sea ice is declining. Gentoos breed almost three weeks later in the season, so in theory were not as exposed to the snow conditions. They have no loyalty to a breeding site, so if they don't like it they move to another, a second advantage. Significantly, gentoos establish themselves in areas of persistent open water. The fact that they are here and have not been here in the last seven hundred years is an indication of true change. This is the beginning of a gentoo and chinstrap region where there were no known gentoos and chinstraps.'

A strong climatic gradient defines the Antarctic Peninsula. To the north, warm, moist and maritime. To the south, the polar environment: cold, dry, continental. Palmer, with its diverse life, is an obvious hinge area where continental runs into maritime. Its status as hinge made it an attractive area for biological studies, and for the establishment of America's third Antarctic base.

The two climates – maritime and continental, warm moist and cold dry polar – shift from season to season, year to year, creating a highly variable environment, extremely sensitive to climate perturbation. This is particularly the case here on the western side, the Western Antarctic Peninsula, a sea-ice-dominated ecosystem where the annual advance and retreat of the ice have a major impact on everything from annual primary production to the breeding success and survival of seabirds. Now, with warming, with sea ice declining in extent and a shorter sea ice season, the maritime climate is impinging on the polar system, slowly displacing the polar specialists. In line with warming, true polar species appear to be in retreat. Fewer and fewer Weddell seals seem to come this far north to pup. Elephant and fur seals, sub-Antarctic animals from the north, the maritime environment, are appearing along with chinstraps and gentoos in large numbers, and increasingly establishing themselves on the Palmer islands. And the Adélies are in rapid decline.

Bill: 'The boundary between northern and southern species, defining the two habitats, is retreating down the peninsula, pushing south. Now it has swept past Palmer.'

When?

Bill: 'My observation is that it has happened in the lifetime of my research in this area.'

It's quiet, intense, in the tent. Bill looks at me.

'This year everything is turned upside down. That is the scene out there. This is truly a year to be here.'

Late in the afternoon some of us are invited on board the tour ship to give presentations about the science on Station. There are cocktails in the lounge with Mr Forrest Mars, who is hosting the high school trip, and a barbecue on the afterdeck with trays of Mars products. Groomed pretty girls and confident boys start partying, dancing on a floor over the swimming-pool in weather that has suddenly cleared from grey wetness to the best we've had. The ice cliffs shine rose and lilac. The last zodiac to leave brings the old hands back to Station, their pockets bulging with candy and chocolate bars.

Two hours later a cyclone belts in. Steffi and I sleep in the corner on the top floor of Bio, exposed to the north. The building shakes in the 50-knot wind, and our door heaves open and shut, our coats stir on their hooks. I lie in my bunk listening to the wonderful racket of roaring wind and sleet slapping on the outside wall next to my head. The wall, two layers of overlapping metal with a three-inch gap of insulation between, is cold to the touch. But the summer can't match winter storms, and this building has stood thirty winters.

In the bunk above, Steffi worries about her research. She hasn't been able to get any samples of the foraminifera she's studying, single-celled protozoa that live here in the cold, shallow water. She needs quality mud from the sea-bed, just the top layer, like the thickened skin on gravy, but all she gets is grabfuls of water or too much sediment. Steffi's good at hustling opportunities. Maggie is diving at 9 a.m. tomorrow and, after finishing her own work, she has offered to collect samples in Steffi's wedge-shaped sampling bag with a bent coathanger for a mouth, the mesh 300 micrometres to let the water out and keep the fora in.

But we wake to the wrong kind of Monday: snow, rain, gales and high seas, with no outside work for scientists. Support staff lost their regular day off yesterday helping with Mr Mars, so they have today instead, relaxing. Scientists have lab work, or notes and reports to catch up on. Senior scientists are tied to office work, the hooks back to the outside world's demands, continuing institutional and administrative commitments wherever you are.

Hugh Ducklow, microbiologist, oceanographer, ecologist, recently appointed head of the Palmer LTER and in Antarctica for the first time: 'I can do my work here, state-side public stuff. It's entirely transparent. Except the phones do not ring.' The institutional base of the Palmer LTER is now at Hugh's home University of Virginia. 'The vision is so important. How do you carry out science when you might not know the answer for a century? The LTER is basic, fundamental research, requiring repeated funding. We are building an edifice for the future. We have real obligations to collect information now, so in fifty years someone can take our data and use it. With some of the areas we don't know what the trade-off will be, but we do our best.' Administering the Palmer LTER takes a lot of time. Hugh: 'It's the first thing I think about in the morning and the last thing at night. But how do you run Bill Fraser?' Hugh permits himself a half-smile: 'Palmer LTER has had a turbulent history. It is a form of peer democracy.'

At Palmer, Hugh is heading a new two-year pilot programme to measure POPs: persistent organic pesticides. An explosion of organic pesticides and industrial chemicals has spread worldwide. POPs get into the atmosphere when the air is warm, not coming out until it is colder, when they reach Earth's extremes. In Antarctica pesticides and chemicals are entering the food web, biomagnifying up the food chain from phytoplankton to krill, on up to penguins, seals and whales: a funnel effect with adverse repercussions on, for example, reproduction. The POPs team want to find out which of the pollutants are in the air and the water, what has built up in the ice and snow, to establish a time-line in the glacial ice, the rate and quantity of the build-up. Warming on the Antarctic Peninsula means an increasing run-off of water from ice and snow into sea water. What is the quantity of persistent organic pesticides currently entering the region's coastal zone and moving on into the base of the food web? This is the first attempt to study POPs in Antarctica.

I go up the track to T5, through heavy snow. Notes need transferring to the computer – my increasingly appalling handwriting, not helped by cold fingers and flapping pages, transmuted to neutral print. It's easy to keep working. The opportunity to be here in Antarctica is so short, every moment counts. Late in the evening an email arrives from my friend Warren in Boston, who knows Antarctica better than most. Stop working, it says. Go down to the bar and talk. So I do, and have one of maintenance man Gary's famous cocktails, and find Chris, quintessen-tial Birder, highly competent but now toast. Burnt out. He's been in Antarctica since August. He'll leave when the *Gould* gets in. To smell the grass, as people here say, to get some green into his secret memory.

We stand outside on the deck. Across the bay there's a rounded dome of ice, chalk-white, the surface weary, splitting like weathered dried wood. Chris understands ice. He tells me the dome is dead. It's our first proper conversation. He has been too busy.

The glacier travelling down from the interior of Anvers Island is leaving this mass stranded, atrophied – tourniqueted at the elbow, the withered arm hanging beyond, no blood flowing into it. Slowly the dome is collapsing into itself – not peeling away and fracturing like the rest of the ice face, but shrinking down, reducing where it stands. Chris says that sometimes the ice in glaciers eddies, like water, and dead ice forms, with the same atrophy effect. There is chalky ice at the head of Arthur

Harbour, formed this way. But the thick dull hump of ice across the bay has been cut off from its mother lode, separated from the glacier, for a different reason. It is resting on land. Bare rocks show around the boundaries where the edges have melted. At some moment in the past the ice connecting it to the main stream of the glacier sagged because there was no land beneath to support it. Now blocks of ice slip and rumble into the sea, which was always there, beneath. The glacier face behind is disintegrating fast and noisily. One day the sea will surge freely between dead ice and living glacier, and there will be a new, small island.

The weathered dome of ice will continue to diminish. Under it a landscape waits, pre-existing, perhaps a headland or two, coves, the dent of valleys. But all will be rock. There will be no trace of soil. All has been scraped bare. The record of life before the ice ground over, before the land bore the weight of this glacier, has been wiped away for ever: the plants that grew here, the animals that lived here, all the fragile, sparse evidence that might have survived has not survived, could not survive, the might of moving ice. Of which this 45-metre depth of dense, chalky ice is the final remnant.

But the ice holds a record, brief, distorted with decay, of regional climate, a vertical archive stacked in annual layers, each carrying a notebook of input – percentages of gases, amount of carbon dioxide, types and quantities of pollutants – specimens of actual air, capsules of ancient atmosphere, time dated, preserved. The speed with which Anvers' ice moves from high piedmont to glacier front isn't known. The age of the ice in the dome is unknown. But all ice is a library of primary sources, and we've only read a few scattered volumes. Ice, once melted, is like manuscripts burned. Records irretrievably lost.

Chris thinks we are looking at what Alaska was like perhaps ten thousand years ago. He says he's just examined a bolus, a ball of regurgitated material, from the nest of a satellite-tagged giant petrel that had flown north up the west coast of Chile, as far as Puerto Mont. The bolus was filled with beech leaves. Chris thinks birds will bring plants and seeds to Antarctica. One day, he says, Antarctica will be like Alaska is now.

12

Sunny day science

Tuesday 15 January

3.30 a.m. The sun glares for the first time through our uncovered bedroom window, riding bright in a blue sky. After breakfast the sky is still clear, the light brilliant, and there is a kind of frenzy to be out. Six satellite transmitters, aerials pointing down, are hanging like tethered black mice taped to a red painted handrail by the snowdrift beyond Bio. A new research project this season will use the transmitters to track where Palmer's Adélies are feeding. But the transmitters aren't ready. They are still capturing their position in the southern hemisphere. They haven't fully adjusted to being at 64° S.

Tracking technology is the current holy grail of seabird research. The ability to measure, the questions that can be asked – all are being revolutionised. Previously unobtainable data can be revealed, information collected almost instantaneously. Huge gaps in understanding can be tackled. Bill's latest waterproof satellite transmitters for Adélies are specially designed to be programmed just before deployment. Technical advances drive weight reduction, and these, weighing in at 58 grams and costing $2,000 each, are reusable and should last a year. A tag's life span is totally driven by its battery capacity. Collecting data barely affects batteries, but transmitting does, so lifespan depends on programming. Bill says he is ahead of everyone, but the advantage won't last: anyone can buy general-purpose animal-tracking tags. Obtaining funding for technology requires agile advocacy skills, and each of his tags is hard-won and precious.

On the print-out in the comms room the system that delivered yesterday's storm can be seen moving across the Antarctic Peninsula, but

another band is building to the left. A line of grey thickness, a misti-
ness, lies over the north-west horizon. It'll tighten up, Bill reckons, but
the dry conditions should hold: 'The weather just manages our lives
every day.'

The Birders are going to the most mountainous of the five inner
study-group islands, Litchfield, to work on Adélies, brown skuas and
giant petrels. It's a kind of mini-environment, according to Bill, exposed
and slippery, pretty brutal in snow and rain. Then they will move to
Humble to attach $200 presence/absence radio transmitters on twenty-
five breeding Adélies, collecting data on length of foraging trips by
recording whether a bird is in the colony, on land or absent. Chick mortality is highest in the first days after hatching, so selected adults must
have one or two chicks at least seven days old. The data accumulates in
a receiver in a hut on the island, then goes to Bill's computer. Collected
every season, length of trips will provide clues to krill availability.

Donna: 'The colonies will be slush. It's going to be a crazy day.'
Blond, blue-eyed, compact, moving with controlled speed, not small
darts, more a fast roll – Donna is doing Heidi's field work while she's
absent on the *Gould* on LTER work, and will take on Chris's work when
he leaves. She's a can-do bustler: tough, highly efficient, eyes always
watching, intolerant of inadequacies. Donna is central to the day-to-day
managing and running of the programme, the logistics, plus ensuring
the integrity of the data, its proper collection. Her objectives are clear:
get through the season, get the work achieved. She does not want to
contend with anything that might delay procedures. Bill: 'Donna is
very strong-willed. She has massive focus. She is completely dedicated.
Everything else is secondary.'

In the Dive Locker Maggie checks wrenches, pliers, a claw-end
hammer and pre-labelled bags, all laid out methodically. Everything is
pre-rigged for suiting up. 'We need so much prep time. It's important
to do a mental check-list, because it's possible to miss something and
leave it behind.' Some gear is personal, such as Maggie's second pair
of socks, made of grey wool (they belonged to her mum). Her orange
vulcanised rubber and latex dry suit, critical barrier between dry and

warm and wet and cold, is borrowed and a bit too big. For Maggie the adrenalin started last night: 'Polar diving is not a natural activity for me yet. I often have the worst problems with my brain – you know you can do this. You'll be fine. For the first fifteen to twenty minutes I think, why was I so worried about that dive? And then, I can't get my hands to stay warm. I do it because that's where my science is, and because I want to. But I have to want to.' Maggie first arrived at Palmer as a 22-year-old postgraduate in 1980 to work on krill, adding, with the need, polar diving. During a pioneering winter cruise in 1986 she observed krill munching phytoplankton off the underside of sea ice, part of a crucial discovery that algae and diatoms wintered embedded in ice, providing food for krill.

The dive plan is discussed with second diver Chris. Plants will be harvested from an experiment set up six weeks ago, looking at factors influencing the production of defensive chemicals in certain plants, and the concrete substrates removed from the sea-bed. Half the results from the experiment are already lost. High winds in December sent a small iceberg bumping along the sea-bed, shoving seven of the substrates, ripping and squashing their racks of plants. Maggie: 'It was carnage down there.' Now the dive is overdue.

Hugh Ducklow and his POPs team load two heavy stainless-steel airtight and watertight bins into a zodiac to collect clean snow from Torgersen. Collecting samples is physically demanding. 'Everything is heavy. Everything takes time.' This morning Hugh has added several tiny bottles to his pockets. He wants to collect an array of the little microbes associated with the snow algae that stain snow a beautiful red. Even here in Antarctica bacterial activity is the foundation of the whole community. Bacteria are all around, scavengers, waiting for the leavings.

Steffi holds the stake boat in a high swell from yesterday's storms, waiting for the divers. Antarctic science involves a lot of waiting. An Adélie pops out of the sea, slim and clean, just as Maggie and Chris trudge heavily to the landing rocks. Strapped into the bags and tank of the buoyancy compensator, hung about with weights, small computer dangling in front, three pairs of gloves, three head covers, weight almost doubled, Maggie looks like a burdened beetle with a heavy carapace. Suffering for science, she says. But what you see in the water makes up.

The boat stops beyond the point, and the divers disappear. The zodiac bucks, and six penguins porpoise in and out of the water, in poignant contrast to the humans feeling around, unseen, below.

The dive, with ten minutes added on for Steffi's work, lasts Maggie's maximum: forty-five minutes. Being a mechanic unbolting substrates under water isn't easy, she says, and she starts packing the harvested plants in equal quantities of dry and wet ice to ship back to the University of Florida. 'This experiment will only be repeated if there is a measurable yield. Something of value. If not, there might be a better way to get at the questions, to try and get an answer. Trial and error.' She pushes back a wisp of hair.

Is there an age limit for divers? 'No. As long as I pass the medical. Being young is no great value.'

In the aquarium room, with its stained concrete floor and wide, shallow tanks, high-energy Amy, postgraduate student, strains phytoplankton. The smell is nauseous. This morning's collected snow melts in the 60-litre steel bins. Steffi sits in the corner of a lab on a round stool working on her newly acquired scoops of sea-bed, leaning forward, eyes almost built into the eyepieces of her microscope, ears occupied by the headphones of her music system. Human as extension of technology. Looking through the microscope, she sees nothing but sediment particles and then, suddenly, a foram: tiny, or tinier. Steffi: 'This is all I'm working on. I don't see anything else. The naturalist approach has been replaced by tunnel vision, amateurs replaced by scientists, who adopted tunnel vision once they had to earn a living from their science.'

Hugh Ducklow, hands in disposable surgical gloves, works at a benchtop covered with white paper, with 'Caution Radioactive Material' written in purple on the boundary of yellow tape. He's preparing samples of sea water taken from beyond the promontory to determine the growth of bacteria, 'spinning and sucking': spinning bacteria against the side of a little phial in a small centrifuge and sucking out the sea water by vacuum, then adding tiny amounts of radioactivity to their cells, to label their DNA. Packed in ice, the bacteria are incubated in the aquarium room. The process has been miniaturised to minimise waste: radioactive water can't be thrown into the sea or got rid of in Antarctica.

The mix of hand-done and high technology is unexpected. Hugh: 'I really like doing this. It's where the data come from. I have to be responsible for each sample because every sample has to be equally good. It helps to keep my perspective. If I stop having contact with what I do, I'd rather have another job. But we have to recognise there is error, and not just human error – inherent error. No measuring system is perfect. All precision gets better. I need to have a sense of what the work is, so that when I see my students, I know how hard it is.'

Chris, the second diver, tells me he'd never worn a dry suit till he came to Palmer, or done any polar diving. 'It's very, very hard.' He dives to collect small bits of brown algae, using a dive knife to cut off a few centimetres, putting them in a mesh bag. It's 'marine natural products chemistry': very new, needing scuba techniques to study small things. Before, you could drop a hook over the side and get fragments of big things. Now there are things you can get only if you go there. 'The ocean is like the rainforest – so big, so diverse. It needs protecting. It's a frontier: undiscovered chemicals must exist. If I find a new chemical, I will find a way to synthesise it.' Then, friendly brown eyes, long brown ponytail, 'Sponges will hide from me … And I like diving in Australia. It's warm.'

Early in the evening a Lynx helicopter settles on the glacier edge, bringing the new Commander of the British ice patrol vessel on the Antarctic Peninsula for a courtesy visit. Last time at Palmer I spent happy information packed weeks on HMS *Endurance*. The visit is deeply nostalgic. For them five months at sea means the pleasure of new conversations. Dave Bresnahan passes on a request from Bill for a series of aerial photographs of the islands, and the state of the glacier. They climb back aboard with Cara's camera and make a special flight, in excellent light. Then are offered the usual coffee and cookies in the bar.

Bill, Donna and Chris rush back to do 'gash' and rush out again to flush the stomachs of penguins. Bill has decided to do diet samples, using an Adélie colony on the control side of Torgersen.

Today's sunshine has visibly relaxed people. Jeff, the boating co-ordinator, talks about his passion for Antarctica. He wanted to come so much that he put in eighteen applications – to do anything. His first season at Palmer continued through the winter. 'Winter people are different. They are here for other reasons.' But he got through it and decided to stay. His mum wrote: 'Are you happy?' 'Yes.' 'That's all right then.'

Jeff: 'I love being here. Having responsibility to solve problems with what I have.' Jeff comes from mile-high Casper, Wyoming, on the North Platte River, colonised by Germans and Scandinavians, now 50,000 people, 1,200 miles from the nearest sea, snow in winter, hot in summer. Brett also comes from Casper, and so does Mongo, the FEM (fuel, electricity, machines) co-ordinator. I think it's the immigrant challenge, coping with a new continent.

There's talk about how the outside world doesn't understand Antarctica. One of the technicians: 'My parents kept saying, "Oh, you are going to the south pole!" Whatever I told them. Then when I finally did go the the south pole, they said, "he's in the Arctic".'

A general assistant: 'I sent wonderful pictures of Antarctica home to a very good friend, carefully selected. Back came the reply. "Thank you! Perhaps I could send you pictures of Oklahoma, and you could send me more pictures of Alaska".'

During the evening Bob emails everyone about tomorrow's tour ship, due after lunch with ninety-two visitors. On Thursday at 8 a.m. a monster luxury liner with over a thousand passengers will moor offshore. On Saturday there's another ship, with eighty-six passengers.

The Birders return to Station at 10.30 p.m. It's too late to process the Adélie stomach samples in the lab, so they put them in the freezer. Day over.

13

Giant petrels

Thursday 17 to Sunday 20 January

Wet, as usual. Grey, as usual. The wind is getting up but it has been building to getting up since lunch. Anywhere else this would be winter and I'd be inside. But we are huddled in a dripping zodiac coming ashore on Humble Island. *Endurance* is hosting people from Station, crew members are visiting Palmer, and Bill, Donna and Chris are giving the Commander and the Marine Commander a special tour.

The shingly slope is vile to walk on, sharp stones slippery with excreta, running with rivulets of brown-grey water. Adélie penguins stand with their backs to the northerly wind looking profoundly miserable, streaked with red mud as though a wide paintbrush has been brandished down their fronts. They slide down rock faces, claws scrabbling on the wet surfaces, try to get back up, and slide down again. Dead birds lie on the ground, bones and feathers squashed: a bit like roadkill. The smell is deeply pungent. The chicks stand in small, bedraggled, disconsolate groups of three or four, beginning to crèche: not a good sign. They are too young, still at the grey, bundly stage. But both parents are starting to go off together and leaving them. Small chicks stand with their heads bent forward in the posture of trying to get under an adult. Except there are no adults.

But the visitors hardly pause to look. There's no photo opportunity in muddy birds. A pair of skuas swoop out of the grey rain screeching their high warning call, defending a territory that this season includes our path. Ahead, elephant seals lie packed together in strong-smelling puddles, great somnolent steamy heaps of moulting hide and hair. They glare with bloodshot brown eyes and gape their pink mouths wide, roaring and rumbling, then belch and scratch. And flop back down and doze. Three years ago the elephant seals occupied the saucer-shaped depression between the higher rocks. Now they lie all through Colony 2. A male elephant seal uses up the same space as four nests. Heave along, heavy body scraping over the clinking pebbles, and the destruction is like a truck driving through market stalls. They can crush without noticing. There's a racket. An elephant seal rolls a bit, a small grey scruffy tangled heap underneath scrabbles itself together, and a just-missed-being-squashed chick creeps hurriedly away.

At the top of the island the landscape suddenly changes. It's as though we've passed through the cheap accommodation on the lower slopes to reach the clear, clean uplands, a rolling plateau of moss and grass and no mud. Nests of pebbles and moss rise discreetly among silver-grey ridges of rock. The owners sit facing into the wind. This is Donna's territory, her major study site of southern giant petrels. In increasing rain, and wind now gusting to 50 knots, Donna performs. She walks up to a nest, feels under the large male sitting so still and lifts out a small chick, seven days old, warm body covered by fine white down, alert little eyes looking around over big pink bill. She cradles the chick in her hands then replaces it under the male who shifts, lifts, adjusts, tucking it back. The chick's head pokes out from under his brown-grey feathers. The visitors are enchanted, lying full length on the ground photographing. Donna smiles her wide, secret smile. As we move past nests she gives a commentary – the bird to the right is a first breeder, so don't go near her; further over is a skittish male, avoid him. She brings us an even newer chick from under a male and asks us to bury our noses in the soft down and say what it smells like. I'm reminded of my children's comfort blankets, when they were babies. Another world, not belonging to this. Donna tries holding the male in one arm but his wings stretch out to balance in the wind, so she puts him back.

The gusts are so fierce that the birds noticeably shake and shudder on their nests. A young bird comes in to land, wobbling with each step in the wind, wings held out, as it almost dances across the rocks, transferring weight from air to island. Another stands facing into the wind, contemplating the opposite manoeuvre. It steps forward as though walking on water, pauses, then runs four or five steps – and it's in the air. I think of hang gliders on the summit of a Swiss mountain this last European summer: four or five steps, running down the slope, then out into the updraft.

Giant petrels are the only bird population at Palmer not to have had serious problems early this season. Their nests, built in north-facing areas, were severely wind-blown, and so avoided the build-up of snow. Data on all Palmer giant petrel sites – who's breeding, who isn't – is collected three times early every season in Donna's exhaustive 'Worldwide Giant Petrel Rounds' Demographic studies are time-consuming. 'We'll all be dead by the time we get really interesting data', Bill comments drily. 'This is a bird that lives a long time. The work will take up to two research lives.'

It's late. Bill radios back to Station, and we go down the slippery slopes past the poor disinherited penguins, into the wet zodiac, and out across the salt-spray wind-chopped sea, to another reception of more tea, coffee and brownies in the lounge.

Bill has been speculating how the reduced numbers of Adélie chicks will conflict with the growth rate of the giant petrel chicks. Wide-ranging, far-flying surface feeders, southern giant petrels don't have territories. They have the globe, foraging eclectically, killing and scavenging over incomprehensibly large areas, indefensible. Their only need is to defend their nests. The tip of a giant petrel's 10-inch long, heavy bill is powerful enough to rip up the skin of a dead whale, but they also eat fish, single krill and young penguins. So the loss of a local resource – in this case, Adélie chicks – may or may not impact them. Bill: 'We don't know how much the parents depend on local resources. There are roughly the same number of giant petrels here as last season, but only one third as many Adélie fledglings available, so there's a bit of a bottleneck. All those pairs are focusing on the same very limited resource. Satellite transmitters tracking the foraging trips of breeding giant petrels from Humble have shown widely differing results over the years – long trips south to the ice edge, north to the coast of Chile, and short local trips. Using satellite transmitters on selected adults we can measure what is happening in real time, set daily information against our long-term database.' But Bill has bet Donna a new horse that this year her giant petrel chicks will be affected. The bet will be decided by data.

Here at Palmer giant petrels feed their chicks on Adélie fledglings at the precise moment the fledglings leave the land and enter the sea. At this vulnerable moment, as the fledglings bob around, too light to get below the surface, giant petrels sit in the water picking them off. But they don't have a behaviour of taking penguin chicks on the ground, although further north in the peninsula, they do. Will they switch this season, and become land predators? Bill: 'Giant petrels are smart. If they learn to do this here, they will wipe out the Adélie colonies. We will have to wait and see what happens. We are holding our breath about all the possibilities. We have never experienced a season like this, with these unknowns floating around. The key factor is to be out in the field, observing what is happening. We can weigh our observations against the tremendous data we possess. Plus add the satellite data. We will be able to say "it is thus."'

Giant petrels can aim a stream of foul-smelling regurgitated oil at an intruder, with accuracy, which has earned them the name 'stinkers'. It's an expensive defence strategy, a week's worth of food, and getting doused is proof of invasion. In the past, giant petrel researchers often burned their clothes after finishing work. Dick Laws, the eminent British biologist, described to me how as young scientists they towed their clothes through the sea behind their boat on the way back to base, after working with the birds, in an attempt to get rid of the smell. I asked Peter Hooper how at Base N during 1955–58 they ringed young giant petrels. 'We held out a bamboo at them and they regurgitated. When their stomachs were empty we could work with them.' Giant petrels are vulnerable to skuas. The colony nearest the Base N hut was adjacent to a skua colony. In 1956 it was reported as 'already depleted through unavoidable disturbance during the nesting season.' Birds flew off their nests and skuas swooped on the eggs.

When David Parmelee arrived at Palmer he included giant petrels in his base-line demographic monitoring of local birds. Nests were visited three times, five days apart, to record band numbers of adult birds, and fledgling chicks were banded. Handling 'resulted in fear responses and increased heart-rate and temperature.' After a female died of 'heat exhaustion' during a procedure, handling was limited to twenty minutes. Painful wounds inflicted by the 'strong, hooked bill' were described, and adults 'viciously defended' even large chicks. But Parmelee's bands only gave a record of who was there. The census didn't establish whether birds at a nest site were pairs, nor was the sex noted.

Donna began visiting the giant petrels on Humble during the 1992–93 season, coming every day. She was in her mid twenties, Bill was the PI, Principal Investigator of the seabird work, and she had only joined the season before as a temporary field assistant after three summers doing initiation snow-shovelling at the Pole, followed by support jobs at Palmer. Her degree was in zoology and biology, she had part-time experience in a zoo, a proven ability to handle animals, and a passion to work with wildlife. Hey, are you hiring, she said to Bill on arrival at Palmer. No – I'm all set, said Bill. Donna: Keep my name in the running. Bill: Yes: stay in touch. Donna laughs. 'We do this all the time. We get up to a hundred requests a year to join us'. But she was lucky. A field worker was released from contract, and Donna was there. A large man, he fell through the ice; on being rescued, his back-pack turned out to be filled with sweet biscuits. He had been hired on the basis of his CV. Bringing Donna in was very much in Bill's tradition of recruiting. Try people – if they can do it, use them.

Up on the Humble giant petrel colony Donna checked who was in each nest, noting which of a pair was male, which female, creating her own learning curve. 'I was struggling for my own foothold. I wanted something to put my own interest into. No one was going to fight me for these.' Bill steered her away from other species. Skuas had breeding failures, then a boom year. Boom, bust, with a lot of bust and some boom. Cormorants were full of external parasites – fleas, ticks, lice, mites, worms. If she ever fell into a puddle of cormorant muck Bill threatened to tow her home behind the zodiac. Giant petrels were scavengers but they were a cleaner species – only ticks, and feather mites. He gave her permission: she could run with giant petrels, 'geeps', in Palmer parlance. Donna: 'I was very green in handling birds. I didn't know how to control them. Birds are unpredictable. And the penguin work was baffling. I didn't understand why Bill was doing it. I had no idea of its potential.'

But on Humble Donna found ways to recognise and remember individuals: head colour – a hat of brownish feathers, or white; the huge variations in eye colour – yellow, deep brown, speckled; chest feathers – spotted, or speckled; the colour of the back. She observed the birds' responses, watching for signs of stress, determining each bird's comfort level. 'I don't remember how the first step was initiated. But I wanted to do more with them.' For two seasons inventory-only work was done. By the 1994–95 season she was ready to begin studies of chick growth, which meant regular handling. So far, she'd never attempted to handle a bird.

Donna went to Humble on her own and tried picking up chicks, 'working out rules. If I pain them I'll stop, if there's a sign that they can't deal with it, I'll stop. Then that's that.' Every day, Bill asked her if she'd been gacked on. No. None of the birds panicked. Bill's response to giant petrels was based on his experience working with Parmelee. Donna: 'It was standard: they're hard to work with. They will gack on you. They're rotten things.' She did it for a week. From then on, everything 'was onwards and upwards.'

New Year's Eve, Palmer, 1998: everyone was celebrating up on the old helipad, bottles chilling in holes bored into a ceremonial sculpture of glacier ice. I'd been at Palmer for nearly two weeks, busy days out on the islands working with field assistants Matt and Pete – knowledgeable Birders, generous with time and explanations. Bill and Donna had just arrived.

In 1984, under Parmelee, six Humble giant petrel males brooding chicks were fitted with back-pack harnesses, each holding a 200-gram local transmitter, possibly the first experiments in attaching solar-powered transmitters to seabirds. One lasted a week, one six weeks, the longest nine weeks. Each time the birds disappeared, untraced. Now Donna was ready to start her own programme monitoring the foraging patterns of breeding giant petrels with four transmitters, paid for by filleting money from another budget. Data would be picked up by satellites in low polar orbit, 160 to 240 kilometres up. Parmelee's telemetry researcher had given her advice. 'You need tape. A pillowcase to put over the bird. Three or four people ready. It's going to be a wrestling match.' Donna: 'He was so pleased someone was taking up the challenge. I pick brains. I ask questions. I don't want to skulk around. But these birds had some level of acceptance. They would accept my presence. I was really worried.'

On New Year's Day, 1999, Donna and Bill planned with Matt and Pete how to put the black transmitter box with its aerial on to a giant petrel, designing a 'feather sandwich.' Lay two layers of back feathers on top of a piece of tape sticky side up. Put the transmitter on top of the feathers and tie with cable tie for added security. Put a piece of tape sticky side down over the box, overlapping a little bit each side, half an inch of sticky to sticky. I still have my example, made by Donna to demonstrate: feathers of torn white paper, the transmitter two packets of Saltine crackers.

Donna was ready for battle. 'I was dreadfully nervous. I went out with a pillowcase from Bio, a fleece hat (all my hats smell of geep. I smell of geep.) I got there and thought, who is going to let me do this and who is not? Matt and I approached together. The best place for the transmitter is the middle of the back, a curve in flight, so aerodynamically sound. If it's too far in front it's a bother to the bird. Too far behind, and they can pick at it. And it worked. The most restraint we ever had to employ is the fleece hat.'

It sounds so straightforward. And in a sense it was. Visited every second day, the Humble giant petrels were accustomed to her; habituated. The four transmitters were attached on 3 January 1999. A month later I went to Humble with Donna while she checked a transmitter on a female. We knelt by the nest and she gave the female her glove to look at, and get broody with. The male was given her notebook. She took the turkey-sized chick off the nest, and asked me to sit on a rock next to the nest and hold it. The chick did not move in my arms. Soft grey webbed feet tucked up, wings tucked in, its body deep inside the softest, fine down radiating warmth. Its little eyes gazed above its bill. The parents fussed with the glove and notebook, nibble pecked Donna a bit, and then the female tried to brood her; Donna adjusted the 9-inch aerial, and we put the chick back. We wore sunglasses, the only precaution: they can go for eyes.

I had never handled a wild animal in my life, except a duckling briefly scooped out of a dawn river, punting one May morning in my ball dress, a long time ago. I'd most certainly never been this close to a giant petrel. The giant petrels on Humble accepted Donna. She trusted me, I trusted her.

Giant petrels were not even divided into differing species with the southerns, *Macronectes giganteus* (the subject of Donna's study), and the northerns, *Macronectes halli* – until 1965. Even now, Donna tells me, giant petrels are seen as second class. 'The sewage and offal bird, the trash bird. If you are a penguin scientist, there's a penguin bible. Ainley. Stonehouse. But a giant petrel bible? No. I'm a curiosity at petrel and albatross conferences. Partly, it's albatross snobbery. Somehow the hackles come through their shirt collars.'

Giant petrels are studied at other sites within the Antarctic Peninsula region. 'There's been suggestions that we all work together on everything. It's everyone's project. But I'm the one who has the most at stake and the most experience in giant petrel work. People say when they hear what I do, or see the photographs, "that can't be. It's just not possible." If something takes your fancy, dive into it.'

On Sunday we have been at Palmer two weeks. Less than two days have been free of snow, sleet or rain. Bill: 'It's been the most miserable January in terms of birds, and ability to get around. Adélie chicks are constantly wet. There's a large number of very small chicks, and they are noticeably beginning to lose ground. A proportion is just not going to make it. They haven't been abandoned, but they've been forced into crèches, both adults feeding at sea. They are too young for it, little tiny chicks. It's a bleak sight.'

The impact on me has been minimal time in the field. When it's wet it's dangerous. Any easing and the Birders go flat out, trying to cram in as much as possible. Either way, they are running to catch up. Donna, grabbing coffee in the galley: 'We are five to seven days behind in our schedules.'

The wind shoves as I walk up to T5, sleet stings my skin. The ground is running with water, big puddles everywhere. I can smell wind and sea. It's just not Antarctica. But to most station staff the weather doesn't matter. It's winter at home in the States; this is just like more of the same but with long daylight and not as cold. My mother sends me a message from Adelaide, on the South Australian coast, that it's 100°F, a hot summer weekend. But it's possible to stay inside an American mindset here in Antarctica, with no particular adjustment to the southern hemisphere. A real gear shift can be required to accept a reversal of the seasons.

Monday the 21st is Martin Luther King Day and we are asked to reflect on his work. Bill says I can watch the Adélie diet sampling. If the weather is OK. Mid-afternoon he calls from Norsel Point: 'It's difficult to get around.' Minimum words conveying coded information. I'm learning to remember that everyone on Station can listen in to all conversations. At ten past five he calls again: 'Be ready in twenty minutes.'

The weather has cleared sufficiently.

14

Meat and two veg

Monday 21 January

6 p.m. The control side of Torgersen facing south to the open ocean. It's taking three practised adults to get one Adélie penguin to vomit its last meal.

Chris goes first. He ranges towards the slight rise leading down to the sea, long-handled black net held ready. Two or three penguins approach up the steep trail from the snow-covered beach, dark polished stones embedded in pinky-brown liquid. Chris quips, 'First contestant?' – walking slowly, then sprinting as the penguins scatter. He scoops up a bird, thrusts his hand into the net and pulls it out at the base of one flipper. The penguin scrabbles the air with both feet, twists its body, jerks its head from side to side, squawking. Chris feels the stomach to see if it's loaded with enough food. It is. So the contestant becomes Penguin 1. The bad luck of being in the right place at the wrong time. Or the wrong place at the right time.

Distant sounds through a thin sea mist of surf on the rocks, quiet clucking from two colonies unseen down the slope at our backs. Otherwise, Antarctic quiet.

The birds are returning from feeding along an established trail used by penguins for centuries, leading to the colonies, both reasonably successful. As we came through, Bill reckoned the chicks looked healthy enough. They are at the standing-straight, scruffy grey, bulging-bag stage, not the leaning-forward, 'I'd rather be under something' stage.

Penguin 1 dangles between Chris's gripping hands while Donna measures the length and depth of the culmen – the bill – to determine gender. It's a female. A washed-up, sea-bleached wooden plank scavenged years ago has been placed flat over the rocks with a yellow plastic cushion in the centre. Chris kneels down, the penguin gripped between his long legs, its flippers along his thighs, its small head point-ing forward, two feet bright pink from swimming poking out behind and firmly held by Donna crouching at the back.

Chris prises open the strong bill. He points the head up and inserts the nozzle of a fuel transfer pump dipped in olive oil, sliding it down the penguin's throat, easing it up and down until it meets the stomach contents. Bill kneels in front of the plank, winding the handle of the orange plastic hose reel, slowly, evenly, pumping in lukewarm water half-fresh half-salt from a 3-gallon plastic container. Suddenly the penguin gags and squirms. Chris starts up. He removes the tube, holds the beak closed, repositions the bird under one arm against his ribs, with Donna adroitly transferring her grip round Chris's left leg to keep hold of the penguin's feet, and – fast – the penguin is upended over a bucket and squeezed from the bottom of the stomach with slow, steady pressure as the last meal is regurgitated in large gobbets. Chris keeps the beak open with his fingers, hooking out any food hanging around inside the throat.

They all bend over the bucket with interest. 'Oh, fish!' The bucket-load is strained through a kitchen sieve, the lump of contents weighed and put in a clear ziploc bag labelled '1'. The bottom of the bucket is examined for squid beaks and fish heads. And the process is repeated. Penguin 1 protests and struggles; in goes the lubed-up tube; she gags; there's the rush to get her the right way upside down over a second bucket. Out through the beak pour the water and the 'dig' – the partly digested contents of earlier meals. Again, an assisted vomit. It's fast, efficient, methodical, practised. I'd thought 'dig' meant 'dig deeper', but it's short for 'digested'.

Another interested cluster around the bucket to examine the con-tents. They are sieved and weighed, and the decision is made to repeat the process. And to repeat it a fourth time, because the contents are still coming. Somewhat lighter than before, Penguin 1, vigorously hitting out, is thrust head-first into a large blue weighing bag. The scale lunges, but at the first moment of stillness the weight is called, and recorded. Chris dives his hand in, grabs the penguin at the base of a flipper and Donna marks her on chest, neck and head with slashes of yellow and green from a non-toxic high-visibility waterproof marker used to mark cattle and sheep. $0.95 a sticky fat crayon, bought in Montana. Penguin 1 now bears proof of lavage, thus avoiding any possibility of another.

Chris: 'Where did I get it from?' Donna points. The bird stands, goes down on her belly, then she's up and running back towards the beach. Where she sits on the snow, facing the sea.

Donna's turn. There's no likely candidate. A penguin approaches and is caught, but turns out to be 'empty', a non-breeder. Down on the beach a scattering of penguins sit on the snow, or stand by the water. But penguins are only coming up the rise singly, or in twos and threes. After much thought and some agonising, Bill has decided to carry out the diet sampling by capturing birds as they come up the track, rather than going into the colonies. The theory is that penguins returning from sea in the early evening will be breeders, stomachs full with food for their chicks. The protocol requires a diet sample from each of five breeders, each week in January and February. The problem, this difficult year, is to find sufficient breeders. Hundreds of penguins normally stream up from the beach at this place. Now there's only a trickle. A limited diet sampling will also be done on Humble in an effort to maintain protocols as intact as possible.

The sea mist rolls in, denser. We can hear a zodiac from station hovering off-shore. They know we are working on Torgersen. But Donna has achieved a bird, and it's full. She briefly puts her hands over the penguin's face, then holds the body tight. Perhaps it helps calm the bird, but it seems more passive. Jobs rotate, and Bill is now in the backside position, holding the penguin's feet, while Chris winds the hose. Watching from the side, I see the penguin's eye, staring, wings flapping as the nozzle is inserted. Bill: it's the most stressful work I do.

A patter of feet on stones from passing penguins. Three kneeling humans in a low line, strangely small on the empty plateau, intersecting the plank at right angles. Unexpected imagery: the penguin as martyr. I imagine a painting by Stanley Spencer, the early twentieth-century English realist, with his domestic settings of Christian visions.

Skuas have gathered, sitting at a distance. Bill: The browns fly across from Litchfield and see off the south polars. They know what is happening.

Penguin 2 is regurgitating dig into the second bucket, the pinky-brown of krill – the colour of the guano putty between the stones of the penguin track, the stains on penguin chests and the fluff of the chicks. Between each event the penguin's beak, the hose and nozzle are wiped with green towels that look remarkably similar to our green bathroom towels.

It's Bill's turn. He sees two penguins approaching in the distance. 'We do love it when they volunteer.' He speculates that one is a male, goes for it, but catches the third female of the evening. Despite being gripped between Bill's thighs, this is a vigorous bird. The first regurgitation is a mix of fresh and digested fish. The third regurgitation is so productive that some spills. Donna checks a sieve-ful of dig, looking like the left-overs from soup-making. She puts the sieve contents on a flat rock at a distance, and the brown skuas fight over a large quantity of penguin undersea labour.

The sieve isn't big enough for the stomach contents of Penguin 4 – at last a male. He's been eating fish. It doesn't seem there's anything wrong with the feeding conditions, Bill comments; late breeding and heavy snow have done the chicks in. Penguin 4 runs fast towards the nests as soon as he's been marked with coloured stripes and put down. The next penguin, caught by Donna, turns out to have ticks. Our world is enclosed in silent white mist. Torgersen is only 500 metres across, maximum height 17 metres above sea level. But this evening we could be at the end of the Earth.

A penguin sneaks past looking sideways, eyeing the activity. After the first regurge they decide to release the tick-stressed bird. Bill removes an engorged tick and squashes it between stones. So a sixth penguin has to be caught. While it's being processed, a penguin comes close and looks. 'Its lucky day', comments Chris dryly.

We walk back to the zodiac down the narrow penguin track, leaving the wooden plank, yellow plastic cushion and penguin net ready for next week's offerings. We carry the plastic buckets and water container, the orange winder with tubes, a rucksack with the marker crayons, one ziploc bag of digested penguin food and five ziploc bags of fresh, the food out first, the penguin's most recent meal.

7.30 p.m. Donna rinses the samples and leaves them to drain over buckets on the deck outside the tent, secured with more upended buckets, and a piece of wood, against foraging sheathbills. Wendy has kept our dinner on plates, and there's a pause for a quick meal before the lab work begins. Diet samples have to be dealt with immediately. Last week's were put in the freezer because the Birders got back to station too late for lab work. So these are the first diet samples to be processed this season. Five penguins have given up the contents of their stomachs so that the size and age class of their prey can be determined and noted according to the Convention for Conservation of Antarctic Marine Living Resources (CCAMLR) protocol: the purpose of the whole exercise.

In the lab the smell of incense wafts above the black-bottomed sorting trays. 'Do not attempt to do this with a hangover', run the instructions in the seabird handbook.

A bump of Jameson's is waiting for each of us, glacier ice melting its ancient water into the golden liquid. Bar ice at Palmer comes from fragments of glaciers, sculpted pieces of solid clarity floating among the salty brash ice, scavenged into a zodiac and carried back to Station.

Donna, Chris and Bill bend over the pinky-grey mash sorting and sifting methodically with tweezers. They started work this morning at 7 o'clock in the tent, entering data from the day before. At 11 they left station for the World Wide Giant Petrel rounds at five locations, and reproduction counts at Torgersen and Humble. But even now the mood is mellow, easy joking running across concentrated work. Donna dutifully deals with her tray, the contents of one ziploc bag. Penguin puke is not her favourite. Chris gets all three digs from Penguin 4 combined, a demoralising heap of mush kept because it contained so much fish. Ruefully: 'It was loaded to the gills.'

Bill bends over a bucket looking for otoliths — small disc-shaped whitish bones positioned one each side of a fish skull to assist balance. They sink to the bottom of the rinsing water like paydirt in a gold-miner's panning dish. He sorts through his tray removing any fish heads for dissection to extract the otoliths. The fish Adélies catch are small, so their otoliths are tiny clouded crystals the size of a full stop or a fleck of dandruff. Most fish are in bits, although he finds a complete ice fish one and a half inches long. All otoliths collected from each penguin are placed in small round black wells inside a four-well container, labelled with the stomach contents number. Otoliths are fingerprints, keys, revealing the age, species and even weight of the fish they came from. An expert in the US will identify each, and the statistics will be added to the database later.

But the krill hit the database in the lab, now. Broken krill, and bits of krill, are set aside because they can't be measured. Fifty fresh intact krill must be selected from each tray as a representative subsample to measure, sex and record on prepared Krill Length Frequency Tally Sheets. Each intact krill is picked up with tweezers, placed along a small ruler and measured from the leading edge of the bulging round black eye to the tip of the tail (telson). Size is noted on paper marked up with size categories 1–8 in 5 mm increments between 16 and 65 mm, and a M/F column.

The krill seem to be crushed horizontally across the body, not broken up or torn apart but partially flattened. A penguin doesn't have teeth, but its long, rough tongue has bristles facing backwards so the krill can't come out. It seems that a penguin catches each krill individually in its beak and swallows it whole, head first. The krill come with their own last meal, a blob of phytoplankton. 'Meat and two veg', says Bill. He presumes that both are of use to the penguin. Phytoplankton are the ocean's grass, and krill feed on the massive blooms that proliferate in the long hours of summer daylight. In living krill the phytoplankton show as a light green ball in their stomachs. Now the phytoplankton is fawny-coloured, and the translucence of the living krill has been transformed to the opaqueness of death.

When he first arrived at Palmer, Bill noticed the kelp gulls whacking Adélies as they arrived in the colonies loaded with food for their chicks, forcing them to regurgitate on the ground. So he did the same and scooped up the spill to measure. Diet sampling by imitation. Since 1988 regurgitation has come via stomach lavage. Much Birder field work is shouldered by Donna and the assistants. But Bill says that as Program Leader he makes a point of sharing in diet sampling and the lab processing. Status nevertheless functions, and Bill gets the best boxes of penguin's last meal. He's positively happy. He's finding huge krill: he calculates six years old – the assumed lifespan of krill. Almost all are female, indicated by a red dot on their abdomen. Most are gravid – pregnant. He dissects one to show me the clustered mass of eggs. He finds very few males. Data from LTER researchers on the *Gould* indicate that the krill are further out than usual.

Researchers on the *Gould* gather their krill samples in great nets towed through the ocean from the ship. There's some evidence that a tow net does not take a representative sample: larger krill can move out of the way. Bill argues that it's useful to get krill from the foragers that use it as prey. This way the krill has already been selected. Diet sampling is an essential part of the seabird work.

Heidi and Brett are currently on the *Gould*, diet sampling direct from penguins. There's a moment's contemplation of the two of them, mid-ocean, doing exactly this lab work but on a heaving ship. Chris has been allocated the dig from Penguin 4, in which the only krill are small and mostly disintegrated or in bits, and he reckons finding fifty intact krill will take him hours – but at least he's not on the *Gould*. He has moved up the pecking order.

People wander in from other labs, teasing and criticising. 'Diet sampling is cruel'. Bill retaliates: 'Penguins are cold-blooded killers. They eat krill alive, swallow them whole.' To Steffi: 'I can hear those forams screaming as you pour formalin over them.'

Bill's second tray has few larger krill and more young. He makes a sequence of size for me, working out when each spawned, its year group, with a large female at the top and five smaller to small ranged beneath. Donna tuts at the time wasted. But it's a clear lesson from a natural teacher, and now I can make reasonably good guesses about a krill's age. The tray has no fish, which saves a job.

Rebecca, coming through from her lab: 'Some people are still working, not just playing with their food.' Bill: 'What you have to do is pick each one out and taste it.'

I look at the range of colours lying in the dissecting trays, from pale grey to light pink. My outsider's ignorance has been jolted by the realisation that techniques have been refined to the point where examining the stomach contents of just five penguins per week, for size and sex of krill, and fish species, can give so much data.

Me: If you were writing the recipe book now, would you include these processes? Bill: Yes. Without a doubt.

11 p.m. I leave them to it. Bill: We are criticised if we're ever late for breakfast. People will say, 'those goddam scientists come down for a holiday.' I go outside into the cold night and walk up the hill, to start writing up what I've seen.

Bill: 'It's the most invasive thing we do to a penguin. But it's better than the old way, which involved guns.'

15

A view from the predator's stomach

An Adélie climbs slowly up onto a rock and stands absolutely still. The bite marks where a leopard seal has grabbed, and lost, puncture its back and chest in two precise curves. Blood drips from a ripped left flipper. At Elephant Island, in the South Shetlands, I'd seen a bloodied chin-strap walk in from the surf and stand on the wet grey sand in the same withdrawn, still way. The ocean's vast anonymity breached.

On land a penguin's body swivels upright, head positioned directly above shoulders and backbone. Short legs swing into commission for walking; stiff narrow wings assist balance. But a penguin's prey darts and sways inside the ocean's layers, or grazes the intricate under-surfaces of ice. The ocean is where penguins search and capture, their wings adapted to long-distance swimming, to rapid manoeuvre, dives and climbs and burst of speed. Legs and webbed feet trail efficiently, rudders behind their streamlined bodies. The ocean is where penguins hunt, and are hunted. The ocean requires fitness and luck.

Escape – and safety – for adult Antarctic penguins means being out of the water. Here they can rest, sleep, moult, heal if necessary, partake in or practise for the business of breeding and raising chicks. Which is why humans and dogs, unexpected marauding land mammals, were so lethal. But the ocean is the penguins' medium, its vastness their privacy. And the core of that privacy is still intact. At Palmer fledgling Adélies leave their nesting colonies at around fifty-four days old, and do not return for three or four years. Where do they go? Bill: 'We do not know.'

What is known is that Adélies do not migrate. An estimated 90 per cent of the animal Antarctic biomass stays in Antarctic waters in winter. Unknown millions of crabeater seals. All minke whales. All Adélie penguins. Counter-intuitively, Palmer's Adélies head south at the end of summer towards the pack ice and winter darkness. Bill: 'It's opposite to what people thought. An Adélie takes circa five hundred grams of krill a day in summer, a crabeater around two to three kilos, a minke six to eight hundred kilos. In winter the light disappears. The classical food web shuts down. How does the southern polar winter sustain this vast biomass? They must all still be taking a hell of a lot of krill. What are Adélies eating? Where are they finding it? In particular, what sustains the krill population – and therefore the predator population? Winter remains the great unknown.'

Since October 2000 Bill has been establishing the first, very tentative, continuity in the summer and winter ecology of Adélies on the western side of the Antarctic Peninsula, linking summer season work, via satellite tags on two post-moult Adélies, to winter work south in Marguerite Bay. The tags, attached in April 2001, tracked the two birds half-way through their journey south before the batteries ran out. During the winter of 2001 two cruises on the international GLOBEC programme, set up to focus on how the winter ecologies of the southern ocean operate, gave Bill precious winter access. Diet sampling provided data on Adélies' winter foraging. Their movements were tracked by satellite-tagging, and two time-depth recorders deployed on a female and a male, both breeders, each lasting three weeks. Data travelled up to low orbiting US NOAA weather satellites during sixteen passes per day. Their nautical receivers, run by the French ARGOS system, passed the data down to Toulouse, and on to control centres in the US to become emails. Sitting in his cabin, Bill could log on and see where his penguins were and what they were doing. 'It blew my mind to think I was seeing the data in real time, as it happened.'

Adélies are birds that cannot fly. Their foraging is limited by the distance they can travel and by the presence of their haul-out platform, sea ice, making the open ocean off-limits by default. Bill established the previous winter that foraging is also limited by the amount of daylight available. 'We now know with absolute certainty that Adélies on the Antarctic Peninsula don't feed at night in winter. They sleep on ice floes on their tums, covered if it snows like humps.' They were observed waking thirty to forty-five minutes before first light at 11 a.m., shuffling off to feed in the sea as long as light was available and coming out around 1.30 to 2.30 p.m. to preen before settling down for the night by the time it was pitch black, 2.30 to 3 p.m. To Bill these limitations give Adélies particular significance in the winter food web. 'Where they go to forage in winter could guide us to regions critical for predator survival.'

The tentative summer–winter continuity has carried on through this season's work at Palmer. At the end of summer, in April, two more post-moult Adélies will have satellite tags attached, followed by the final series of GLOBEC cruises next winter, when Bill plans the deployment of six time-depth recorders. 'We went into last winter ignorant. We had no knowledge of what to do. But the birds told us where to focus our work. We saw where they were and where they weren't.' The Adélies seemed to be hanging out in the extremely deep holes in Marguerite Bay, two known polynyas – semi-permanent areas of open water in the pack ice with warmer deeper water – where krill seemed to be congregating. The polynyas last all winter long. 'It became obvious that to sample Adélies' diets properly we had to do more than work off a ship. This next winter we will set up a camp on Avian Island, smack in the middle of a polynya, and diet sample there. Our work will be the same but more focused. The hypothesis remains the same but what basket you put your eggs in changes quickly. Last year we saw our limitations. Now we are better informed.'

Last summer, diet sampling established that Adélies left the Palmer colonies with average weights. But birds sampled during winter in Marguerite Bay were lighter by 500–800 grams than on the two other winter cruises with data that could be used in comparison. Even in August and September 2001 the birds were light. Bill: 'My impression is that the birds arriving at Palmer for the start of the summer season in October were in poor condition.'

This fits with the tentative answers Bill has been giving to my probing about the drastic drop in the numbers of returning Adélies at the start of this season. Fifty per cent no-shows is massive. 'The penguins were there this winter, but not in good condition. I can't help thinking about the possibility that the bad times began in the winter. Whatever happened began to happen after they left Palmer. This winter the penguins might have died en masse. On several occasions we saw giant petrels eating dead penguins. The giant petrels are doing so very well. We didn't predict that. But a massive die-off of penguins on the sea ice would create a favourable environment for them.'

Why would a mass die-off of penguins occur in winter, if it did? Was there less food? Bill: 'Sea ice somehow provides access to prey, directly by attracting prey to the under-surface of the ice, or as a surface platform for the Adélies. The sea ice formed last winter two weeks late: that could have affected the getting of prey. Plus the heavy snowfalls sitting over the surface of the ice shut out the light, delaying diatoms and algae.'

The summer–winter continuity work ties in with Bill's efforts to understand the relationship between Adélies and krill – predator and prey – using data gained from the penguins themselves. A predator's viewpoint. The view from the predator's stomach. Diet sampling reveals with precision what a penguin has just found to eat in the ocean. Bill shows me a draft of his latest article. The hypothesis is that the Adélie does not select certain sizes and ages of krill: it takes a sample of what is there. A cohort of krill stays in roughly a preferred area, and Adélies' feeding relates to these areas. Using his data sets and working with oceanographer and numerical modeller Eileen Hoffmann, a hypothesis has been developed that the sizes and age classes of krill found in sampled Adélie stomachs correlate with krill's life histories.

Krill probably live for five to six years. It is absolutely vital that within that life cycle gravid females can deposit eggs that have a chance to grow and flourish, to replace krill numbers. Spawning occurs from December to February. The eggs drop deep into the ocean, 500 to 700 metres, to hatch in a layer of warmer water, Circumpolar Deep Water, which maintains its identity as it flows around Antarctica. The larvae rise up through the ocean's layers, to feed on algae found under the surface of sea ice in the following winter. Current thinking, and the hypothesis underlying LTER work, is that the under-ice habitat with its communities of microbes is the winter grazing ground for larval krill, playing an important role in winter growth rates and, potentially, survival. A high-ice year – a year with extensive sea ice – provides favourable conditions for a strong krill year, a good recruitment of krill more than a year old.

Bill's consciousness of the decline in strong ice years on the Western Antarctic Peninsula grew out of his research in 1988, which led to the 1992 climate warming paper. High-ice-year cycles began with the winters of 1975, 1980, 1986, 1990 and 1994. Bill has been arguing that if the essential high-ice years stretch out to occur four or even five years apart, this reduces the opportunities for several consecutive strong krill cohorts, limiting the krill population to only one opportunity a generation to regenerate. A larval female hatching, for example, in the strong ice year of 1980 was at the end of her life before the next strong ice year of 1986. Extensive sea ice, providing – according to current hypotheses – conditions favourable to the generation of strong krill year classes, has been rare in the region for most of the last three decades. The same situation has occurred now. Larval females hatching into the last strong ice year of 1995 are at the end of their lives. Bill has been pointing out the risks inherent in the tenuous nature of krill life-expectancy and strong ice years, if his analysis is valid. One gap too long – one miss in the beat of good ice years – and the krill population could crash. Krill underpin vital ecosystems. A crash would have dire effects on all predators depending on this source. In 1986 Palmer's Adélie population fell steeply.

Established krill scientists have spent their working lives on their subject. It is a controversial area to enter. A mass of converging factors impact, relating to ocean currents, observed periodicity in ice cycles, extent and timing of sea ice, water temperature, salinity, the behaviour of polynyas, the life history of krill larvae, the migrations and behaviour of adult krill and the intensity of reproduction of the female. But Bill is a free-thinker. Ideas drive him. His work with Eileen Hoffmann is an attempt to pin ideas with data, using the predator's perspective as evidence. Winter is clearly critical to unravelling the Adélies' demography. Crucially, winter has seen the greatest rise of temperature on the Western Antarctic Peninsula. But for Bill, 'We know so little about the evolution of Adélies in winter, almost everything is speculation.'

16

Dream Island

Tuesday 22 to Friday 25 January

Palmer winds the focus in tight, to the immediately local. Going up the hill to my desk is a decision requiring at least, this season, a coat. Wandering beyond T5 in the Back Yard means signing name, destination and time of departure on the blackboard, signing off on return, but not taking the two-way radio needed if walking up the narrow corridor of permitted access on the glacier. On the glacier the interior of Anvers Island unfolds to the north, untouchable. To the east, the mountain spine of the Antarctic Peninsula with its clustering off-shore islands is a panorama of shifting images.

But Palmer's bonus of local islands gives a rare freedom, and the seabird team has the widest range. The intensity of their investigation relates to distance from Station buildings. Inner islands are vigorously worked, the chance of location consigning their birds to detailed scrutiny. The islands at the edge – the Joubins, Dream and Biscoe Point – can only be visited as and when limited labour allows and the weather permits.

On Tuesday a small weather window opens, and we take it. Twice every season the Birders achieve a broad-brush census of adult Adélies and chicks in selected colonies on Dream. The established chinstrap colonies are censused, and brown skua breeding success recorded. Three years ago at Palmer I spent happy sun-glittered hours at Dream helping with penguin counts and scrambling up to watch brown skua Pair 1 on their bold rock-ledge nest overlooking a broad penguin domain, busy colonies clustered on slopes and ridges, the ice cliffs of Anvers as spectacular back-drop. Dream is big and beautiful, which is why a chance to get there is so lusted after. But it's 9.5 kilometres from Station along the coast towards the open ocean, near the edge of the manageable boating limit. This morning it's a grip-tight, bottom-bouncing-off-the-edge, keep-concentrating ride in vigorous swells.

A penguin thrashed out of its skin lies on the snow, red meat on a white tablecloth. The stony plain, currently a snow plain, is littered with the season's reality: broken eggs, carcasses, dead chicks. The landscape effect lies spread out in front of us, textbook clear. Penguins nesting on high ground scoured of snow by northerly winds are maintaining their nests and chicks. But penguins on low ground have failed.

Up on a rock crest a remnant Adélie colony has just five adults and a single good-size chick, well-looked after. Bill. 'This is the saddest sight you will ever see.' The chick will not, cannot survive. Generation after generation of birds has been recorded at this colony. Fidelity to the breeding site is so extreme among Adélies that the adults will not move to another colony despite the detrimental conditions. They will come back to this exact place, over and over again, until there are no penguins left. In another colony there are only three chicks: one small and helpless on the ground, one mid-fluff, the third standing and biggish. Bill: 'There's no hope for them.'

On top of a single high rock I see an unbearably poignant tableau. A medieval dance of death. One fluffy chick is standing, very still, on its pebble nest, with one adult. A skua stands almost next to them. Waiting. Death openly in attendance.

When Chris and Heidi achieved Dream on 1 December to count occupied nests, they found some of the smaller colonies buried by the fierce storms of the previous three days, birds deep under snow. Now a straight headcount is made of Adélie chicks at the same colonies, to give an indication of survivors. A 1:2 chick count is done, a 'Bill special': how many nests have two robust chicks, how many one only. Bill: 'It's a timeline, a statement of the influence of the environment on breeding.' A clicker is held in each hand, one for one-chick nests, the other for two-chick nests, but this difficult year it's difficult to do, and the results miserable. At the chinstrap colonies active nests, occupied nests and total numbers of chicks are counted. Everyone counts independently, three times, and the results are averaged. Donna moves along a colony counting from the left, sometimes holding up one hand to distinguish boundaries in the shifting mass of birds. Bill teases her. 'When she started working with us, she scored fewer numbers. She's so short, she couldn't see.' Donna tells me what it was like working between two tall men in her first season, urgently jumping up to try and gain height, constantly pleading, 'Tell me what you're doing'. Counting gets addictive. Back in England, I add up sheep in fields.

In the mid-1980s the rock ridges and flat spaces of Dream Island were estimated to hold twelve thousand pairs of Adélies, in vast colonies. Three years ago when I was here, 187 pairs of chinstrap penguins had infiltrated the edges of some Adélie colonies. This year, after the heavy spring snows and deep drifting, the chinstraps are doing fine – colonies busy with parents and chicks, plus the usual losers, scouts and pre-breeders – and pair numbers are up to 251. Adélie numbers have dropped, but only the east colonies are censused, so an overall figure is not known. But some of the larger colonies are holding on semi-well, with concentrations of three or four darkish grey chicks, scruffy and grubby, clustered together, crèching. According to Bill, they look in good enough shape. The weak chicks have probably already been taken by the skuas.

I watch a biggish chick ricocheting around a colony, driven in short, jerking bursts by adults darting pecks. It stops, is pecked, runs, stumbling, stops – is pecked from three sides and manages to push past. It stops with one or two chicks but it's no safe haven – is bundled on, tries a group of four huddled chicks but still the adults dart their open beaks at it, on and on like some ghastly game until it reaches an outside corner of the colony, where an adult accepts it. The chick clicks and taps the adult's beak, the adult bends forward and clicks and taps back. For the next thirty minutes that I can watch, the chick stays reasonably close, the adult seeing off any other bird that approaches. Three chicks huddle together nearby, hardly moving. In the middle of the colony a small chick is being heavily pecked for not being where it should.

It bows its head and receives the pecks. Chicks have at a certain stage to crèche for safety, yet are punished for wandering. Innovative chicks appear to risk death, conformity appears to be rewarded. But for Bill and Donna descriptions of behaviour are not of relevance.

By the time we work our way past the ever-increasing numbers of elephant seals and a few frisky fur seals back to the zodiacs, clouds and mist are rolling across the island and the wind is very cold. It's 4.30 p.m., and we rest against rocks eating our lunch. Four giant petrels were sitting in the snow like Christmas geese when we landed – three of them banded but too far away to read. Giant petrels don't nest on Dream. Now one flies low over our picnicking heads, with a satellite transmitter on its back. Donna monitors approximately 900 giant petrels, and her budget has stretched to six 35-gram transmitters this year. She is ecstatic. It's as though the bird is doing a fly-by.

On the way back to Station with following seas through high swells and wind-blown chop the cold salt spray defeats my Punta Arenas-issue wind trousers and all layers beneath: fleece trousers, long johns and knickers. My hair is wet despite hat and neck-warmer. My gloves are wet. But Donna intends weighing the Humble giant petrel chicks, a task repeated every second day, so we land and stumble on chilled limbs past the Adélie colonies and up between the elephant seals.

I used to feel a real interest in elephant seals. Sunshine must have helped. There's no denying their sheer size: up to 3 tonnes, 5 metres long for a fully grown male. But this year there are so many on Humble, their heavy bodies so intrusive. Lying on top of each other, making their short loud interjections of noise, white stuff squeezing out of their noses, their skins peeling in scrappy old-rug scruff, they seem so peculiarly lacking in form: giant slugs with at one end eyes and a nose and a mouth, at the other a small, divided, flippery tail. Now we wade through their slush and muck, on up to the mossy highlands.

Donna reaches under each adult and removes the chick, measures length of the culmen, calls the figures and hands the chick to Chris, who chooses a bag according to estimated weight – tiny mesh, bigger mesh and big blue – each hanging from its own scale, and calls the weight. Bill notes all figures in the notebook. The chicks don't struggle. It's over so quickly. When Donna broke her thumb on Humble two years ago, she had a scale drawn on the plaster so she could measure chicks against it. Now she moves fast, following a route around her flock of forty pairs, always noticing. This bird has a new mate. That one has left its chick in the nest: it's too young, so she tries to persuade the bird back on. A small chick is wet, and she tucks a few stones into the nest. Younger birds often build nests with a lot of moss, which gets damp; experienced birds build better nests using pebbles, hard to sit on but well-draining. Several birds she does not approach: one is young and new. Donna suspects younger birds see older birds here and so feel secure to nest. Spare nests and sites are available among the rocky knolls and rich mosses, and the colony is increasing.

Birds exhibit a range of responses, some nibbling at her hand, some uttering noises, a few appearing flustered, others passive. Young Adélie chicks stick their heads under a sitting parent and present their backsides, poised to arch a stream of guano efficiently away from the nest. The single giant petrel chicks peek out from under a parent alertly. Donna removes one from beneath an adult on its nest. 'Push your hand right under', she says. I kneel down directly in front of the giant petrel. His feathers are brown-grey, like the rocks. He's bigger than a swan, with a wingspan, if he chose to stretch out his wings, one and a half times my height. Heavy, yellow hooked bill, golden eyes. I take off my glove, check my sunglasses are in place and slide my cold hand in under his body, past the smooth feathers. In the middle I turn my hand. Suddenly I'm inside the brood pouch. Warm, smooth, rounded, a secret living cave. Extraordinary.

Bill looks a bit put out, Donna's geeps are taking all the attention. I say, 'Why don't we feel the brood pouch of an Adélie?' 'The bird would break your wrist before you got there,' says Bill, with a certain grim satisfaction.

A giant petrel fitted with a tag flies over and lands in one of this season's two comparatively successful Adélie colonies on Humble. Bill and Chris lope over and find the big bird shredding an Adélie chick. They think it was already dead. Donna tells me she sees all kinds of things in giant petrel nests: chick eyeballs, squid beaks, found objects. She hands me a penguin chick tongue, narrow and yellow-brown

Two Adélie chicks have followed adults down to the top of the snow-covered beach in a food chase. Bill: 'They'll get caught by skuas.' Three adults appear to be staying with the chicks, perhaps as protection. But if the numbers this season were normal, there would be a lot of adults on the beach, many more chicks and much less danger. So the slide downwards in the penguin population facilitates the killings, accelerates the reduction in numbers.

Last evening I looked at phytoplankton inside krill that had been eaten by an Adélie penguin to feed to its chick. Now I look down the slope at a giant petrel eating an Adélie chick which will be regurgitated to its own chick; which Donna will weigh, noting the increase.

We've been out for eight hours.

Wednesday is Palmer Extravaganza Day, the fortieth birthday of Bob. It begins conventionally enough, with all Station members gathering in honour of our station manager at a surprise breakfast in the lounge, donuts covered in coloured icing and donut holes tossed in sugar, and a present from Susan, the visiting teacher, of fluffy pink earmuffs decorated with tinsel, which he is obliged to wear. Except that Bob's wary eyes and cautious back-against-the-wall stance should have been sufficient indication. And the raucous rooster wake-up calls at 5.40 a.m. through Bio, repeated at 6.30 with screeching bagpipes, because his bedroom had been so cunningly wired.

Bob's day unfolds inexorably punctuated by repeated and exhaustive inventiveness, from his office door taped up with shredded paper and heaped popcorn confined in food wrap, ready to spill as soon as he opened it, to a 20-foot zodiac inserted into his bedroom (uninflated) and then ingeniously inflated so that its bulk fills the space, jamming against the door so he can't – finally – get into his bed. His office is taken over by repetitive lacy bras, a big red salvaged buoy pumped fully up and a black inner tube, ditto. At his birthday dinner, his eighth in Antarctica, he has to endure a swipey kiss from every woman on Station, each daubed with Donna-provided, cheap smeary lipstick.

Thursday is grizzly dull and windy. An elaborate murder game begins, helping to siphon off spare creativity still swilling around after yesterday's birthday. Up in T5 I reckon I'm immune to murder. So immune, I don't know I'm still alive. The Argentinian navy drops by, and seven mellow men in blue boiler suits are invited in for a special pizza lunch, much conversation in Spanish and a Station tour. Their small grey ship – not ice-strengthened, complement of 73 – began life in France during the Second World War as a US patrol frigate.

After a day's slog at my desk, regardless of murderers and weather, I need a walk in the Back Yard. To the side of T5 the land dips down to what was the melt-water lake from the glacier thirty years ago when the Station was built. Runs of overground cable lead to networks of aerial arrays. A few tiny plants colonise rock faces. A few small tents colonise hollows and dips. The Back Yard is why T5 is dirty, according to Orion. It's a wasteland of rubble, glacial silt and rock reduced to the consistency of talc powder by the grinding of ice. It is alien, uninterested. The edge of the ice age. The unsettling edginess of its edge.

On Friday the murder game hots up. Calm snow falls. The number of 'dead folks' steadily increases. Kristin visits T5 for a once-a-month building inspection: safety features, fire and exit notices. 'This is a depressing building. So many projects happen here.' Me: 'It's industrial. I can't even colonise the desk.' She and Orion invite me to come with them to Torgersen. We wander through thin snow on the open-access side. Small groups of Adélies are separated by huge gaps, remnant colonies all on high places, like land after a flood. Several nests are so tall that approaching birds bump their beaks on the piled pebbles: a glut of resources, and no stealing. A pair of (presumably) teenagers do random mating, the male trying to reach over and touch beaks but getting nowhere near relevant bits of anatomy, then hopping down, leaving muddy tread marks on the female's back. Male and female Adélies look identical. Measurement of length and depth of the bill is used to distinguish which is which. When fledglings were banded at Palmer, an attempt was always made to establish sex. But sometimes, observing mating over subsequent years, Birders saw their mistakes: a male, with tread marks.

The chicks are in grey woolly fluff, perhaps half of them crèching. I watch a food chase: two large, rotund chicks running after an adult, shoving, pushing, trying to get past each other, flippers overlapping like a rugby scrum, stumbling, slipping, entreating with high piping cheeps. But chases seem circumscribed, the adult not going far from the colony and taking the chicks back again to the edge. A small-ish colony is being harrassed by a pair of skuas. The adults stand in a line, facing outwards, bills snapping and darting. But at a small, small colony nearby a skua pecks the backside of a chick lying by itself. It does not attempt to move.

Back at Palmer, as we walk up the slope past the boating shed, IT specialist Hugh, lying flat and still in the bottom of a beached zodiac, uncoils like a trained spring and I'm ambushed and murdered. It turns out I'm the last left. It turns out he's ex-army: so I reckon it's an honourable death.

17

Boondoggle Day

Saturday 26 to Monday 28 January

Out in Hero Inlet, Bill turns his zodiac in tight fast circles, checking steering, the motor, fuel levels, before loading gear and people at the new boat ramp. Since achieving Dream on Tuesday rain, snow, wind, snow, rain have denied all fieldwork except the most essential. Now, Saturday 26 January, the weather has switched and the sun is shining, the sea calm. It's the day for Biscoe Point, the most southerly site for seabird work and the furthest – 12 kilometres back towards the ice-clogged corner of Anvers, where the Neumayer Channel opens out into the Bismarck Strait. Like Dream, Biscoe is accessible only in calm, clear weather, and work has to be on an opportunity basis. Biscoe is a Site of Special Scientific Interest (SSSI), and chances to get here are rare. Today's expedition is a farewell trip for field team leader Chris and for Hugh Ducklow, both leaving on the *Gould* on Monday. And there's space for me.

Significant places that are also beautiful have the ability to confer a sense of private ownership – this is my room, my view – when with a moment's reflection you know that you are just one in a long line of participants. But in Antarctica there are so few of us. And chances may never come again. Antarctica is free from many social constraints and structures. It belongs to everyone. Yet there is an irony in this generosity, this neutrality of ownership. It can breed a kind of hunger, an edgy competing, intense levels of wanting, of frustration. In this place of confined communities an economy flourishes, a currency of trading and owing favours. Hierarchies of insider knowledge, of access, of the ability to confer and withhold, insinuate and grow. And the most valuable currency, heavily competed for, using scarce resources, is the treat, the goodie – to Americans, the boondoggle. To British, Australians, New Zealanders, the jolly. Distributing resources is the ultimate power in resource-strapped Antarctica. Jollies breed jealousy.

Bill and Donna are at the top of the Palmer tree. Bill especially has been here so long. They know so much, they come every year. Crucially, they are the conduit to the islands. For many on Station, knowing everything that happens, being inside track, is part of being here. People want to see what the Birders do, experience how they find out. Bill and Donna have friends, people to oblige, support staff needing opportunities. Some people ask repeatedly to accompany the Birders; contacts and status are levered, obligations called in. They attempt to distribute trips evenly, but space in the boats is at a premium, and there are busy schedules to get through.

I hunt sandwich fillings in Debra Jo, the kitchen left-over food refrigerator, named after a model who sent Palmer pin-up photographs. Her minimally clad form used to nestle inside, pre-political correctness, the first thing you saw on opening the door. One summer an official arrived and cleaned up. I fill my regulation water bottle, add layers of clothes, pack my dry bag, choose a good float coat, make a last visit to the loo. Dehydration is a problem in Antarctica, reducing alertness and affecting decision-making, and we are meant to keep drinking. The risk of dehydration is a toss-up with the fuss of peeing on field trips. It's permitted between high and low tide. That's an easy convention on tropical beaches. But here the crucial tide gap isn't all that accessible, and there are a lot of clothes to deal with. An alternative mechanism exists. The American product has a polite name, the Lady Jayne; Australians call theirs a Sanifem Pisaphone. But it needs practice, and somehow, in London, I didn't.

No Birder journey seems to leave out humble Humble. One of the six satellite transmitters failed, the others have been deployed. This morning Bill wants to catch a big male Adélie wearing satellite transmitter 14535. To minimise potential stress Bill has decided to rotate the tags. The wearer of 14535 jumped ashore a few hours ago, ate snow from the snowbank and walked up to Colony 3.0. When Bill locates him, he's standing alone. But by the time Bill is ready he's moved into the centre of the colony, and Bill decides to use a net. By the time he's caught, birds are running, squawking, chicks scattering.

The penguin struggles, head jerking, feet treading air as Bill carries him by both flippers to a group of flat-topped grey rocks and sits, bird across his knees, feet held together, flippers held firm. Donna puts a grey wool sock over its head ('dark is calm for most wildlife'), and Chris begins cutting off the cable-ties holding the black box in place. Except he can't, and has to try three pairs of scissors from Donna's capacious everything-needed-in-every-emergency backpack. Once the box is free from its tape, Chris pulls the feathers away. The lumpish bit left will moult off in March. The penguin is completely still across Bill's knees. I take the chance to touch his pink feet – surprisingly soft, the black back feathers bristly, the breast feathers soft, with a sense of quills beneath. I run my finger down the hard leading edge of the flipper. Hard enough to break a bone, reminds Donna. The little box is checked for signs of damage or wear after its first use. A thin aerial logs presence/absence, revealing data about the length of foraging trips; a thicker aerial records journey data, giving Bill an insight on how far his Adélies travel to forage. Data can only be successfully transmitted when a penguin is out of the water, on ice or dry land. We speculate how deep the box has dived and how often, but this isn't one of the even more expensive dive-depth recorders.

So much information is wanted, so much not known, but every extra specification adds cost and weight.

Released from his week-old package, the penguin is carried back to where he was found and returns, after a brief pause, to his single chick. Bill stands, tall and thin among the penguins, in faded blue poncho top with kangaroo pocket, wind pants, inevitable cap, scanning the colony. Quiet. Still. He wants another male as the next transmitter-carrier, but with slightly younger chicks, to see if there are differences in the sequence of journeys. Suddenly the net is out – a bit like the quick dart of a penguin's neck and beak when aggressing – and a struggling bird is in the mesh. Bill grabs a wing, grabs the other and walks back to the flat-topped rocks. To a reversal of the process I've just watched.

Chris applies two dabs of glue to the penguin's back to create two feather pads. Bill eyes the line and suggests moving the lower block of glue a bit to the right. It's tested for tackiness. Instructing is going on, the usual Birder technique of learning by doing. In three months' time Chris will be three degrees further south, on the first of next winter's GLOBEC cruises, applying $20,000 worth of equipment to Adélies on seabird-abundant Avian Island in Marguerite Bay. The glue is ready for the transmitter. Bill: 'Lift up a good quantity of feathers to make a solid pad.' Tape for securing the box is ready on the side of a plastic bottle of hot water 'It works better if it's hot,' says Donna, and she jostles with Bill for the right to claim the technique. The tape is German, linen-coated, almost indestructible, forming a watertight seal between feathers and transmitter, but it's been cut to the size of previous transmitters and this one is thinner. Careful, concentrated work as the tape is wound around both ends of the feather-pad and transmitter package, the extra cut off, the box tested for stability, cable-ties added for extra security and spare cut off, and the package tested for solidity. I try wriggling it, and there's minimal movement. The newly equipped penguin is deposited back in the colony, shakes himself, twists his neck around and doesn't return to his chicks. Bill worries.

Donna needs to put satellite tag 22967 on a female giant petrel. We cross the lowish land, past the mucky pools, the slippery stones and heaped elephant seals. The seals roar a bit. Penguins squawk. There's no wind, so the skuas targetting Bill and Chris come from all directions. Scrambling to the uplands, Donna discusses with Chris which female to choose. Who's here? Is her mate in attendance? What's her mood? Mood can relate to length of time on the nest, or hunger. A bird accepting of contact might suddenly get annoyed. An annoyed bird recently hit out, biffing Donna's knee. She chooses a female known to be docile, from Nest 10.

Chris kneels and gives the bird a mock egg to brood while Donna eases the pigeon-sized, very young chick out from under and hands it to me. Sitting low beside the nest, I try to keep the chick's feet up and together and its wings in, without squashing. Inside my hands the white down is profoundly soft. After ten minutes the chick begins to cheep. I worry it might be getting too hot, so I adjust my hold and it settles. Donna touches the female, talking to her, touching her big bill, then her back, to see if she minds. Ignoring her chick, and me, the female shifts around in the nest. All the time Donna is making up her mind whether the bird will accept the equipment. Applying the box to the middle of the female's back takes fifteen minutes. The procedure involves no restraint. If the bird is unhappy, she can leave and the instrument will drop off – a disaster. I ask what is in the very dark giant petrel poo. Answer: penguin. There's no doubt that Donna's giant petrels have begun to prey on Bill's Adélie penguins.

The last task on Humble includes everyone; no choice. The pair of brown skuas nesting this year on the seaward side of the path have two chicks. We squat on the ground while the chicks' culmens are measured and weights noted, in a fine rain of little scales, bits of skin and feathers. The frantic parents wheel towards us in high agitation, sweeping in, calling. One wings past Bill and scores a hit on his back. Then both chicks are put down together, exactly where they were found, minimising the fear of their being abandoned or being eaten by other skuas. Bill returns to the Adélie colony to check on the satellite bird. But he has gone.

We boat on to Limitrophe Island, green with grasses and mosses contributed by the kelp gulls that once nested here. Bill shows Hugh and me natural rock pools with clear fresh water, the largest, like a Roman bath in the rock, with tiny shrimps. The pool freezes solid in winter and in summer can dry right out, yet there are still shrimps. Crisp Hugh matter-of-factly puts on surgical gloves to take samples of moss for his POPs project. Donna and Chris need to census the giant petrels nesting on Limitrophe and nearby Hellerman Rocks, so Bill and I go on a mini-expedition scrambling along a steep-sided cliff in the sun and the wind to the sites of abandoned kelp gull nests: small shallow communal hollows on the soft turf, near the edge. Grass grows out from each site like a green benediction. The moss that kelp gulls bring for their nests has grass seeds tucked inside. This is where Bill researched as a student. Kelp gulls overfish an area, collecting smaller and smaller limpets until there are none left and they all migrate to a new territory. Bill shows me where the gulls ate limpets on flat rocks, the regurgitated shells blowing and sliding down the hill into a hollow – a limpet midden. I push my fingers in and find it deep full of thin limpet shells, stacked inside each other.

Then we are back to the boats, zipping up float coats and attaching beaver tails, pulling on neck-warmers, second pairs of gloves, ear protectors and sunglasses, and speeding along the glacier front, two small boats, five happy people, over a silky smooth sea, elliptical inter-leaving of blue and grey colours on the minimal swell. The peninsula mountains are spectacularly clear. The two boats swerve and overtake, and I hold on tight, sit straight and make sure there's space among the bags to drop forward on my knees. I don't want to flip backwards and be fished out of the sea, as happened once to Bill's National Science Foundation boss.

We tie up in a narrow loch ending in pebbly shallows, with silvery sand and bright-coloured seaweed, and climb up to smooth grey rocks for a contented 3.00 p.m. lunch. The glittering splendour of Mount William rises centre-stage. Exactly where John Biscoe landed in 1832 isn't known but I'd like to think this is it. Across the shallows, Adélies nesting on the ridge cackle sporadically. There's time to notice the purest air. Freedom to listen to natural sounds.

Biscoe Point feels a mature, balanced piece of country, not like the unfinished roughness, the angularity, of some of the islands. SSSI des-ignation came because of its swards of Antarctica's two native flow-ering plants, *Colobanthus quitensis* and *Deschampsia antarctica*, growing with relatively well-developed loam beneath. Several hundred species of vascular plants occur in the southern tip of South America, Tierra del Fuego. But only two have succeeded in crossing the 900-kilometre barrier of the Drake. So far, nothing has got established along the

Antarctic Peninsula in probably the last ten thousand years, except the enigmatic little hair grass and pearlwort.

Three years ago Bill brought me a present of a drink of ancient glacier water, caught as it trickled from an ice dome, part of the ice ramp that used to connect the hills and valleys of Biscoe Point to the bulk of Anvers Island. Now, with warming, the dome has almost gone. The hair grass and pearlwort are no longer a rarity but are spreading rapidly. And Biscoe is no longer a point but an island, with a deep-sea strait running behind its back.

After lunch we wade across the shallows of the loch and climb the hill for the Adélie total chick counts. It's a sadly quick job. All the smaller colonies here have failed completely. The larger colonies are severely reduced, with few chicks. In 1984, 2,800 pairs of Adélies were censused. Now the score of chicks is dismal. The penguins, rarely visited, seem more nervous of us, more agitated. Bill says, 'Yes, it's true.'

The two brown skua pairs have failed, and the hybrid skua pair. Palmer has a thousand breeding pairs of south polar skuas, *Catharacta maccormicki*, plus two hundred singles congregating in clubs. But the dominant skuas at Palmer in terms of behaviour are the browns, *Catharacta antarctica*. This season eleven brown skua pairs are being monitored. Brown skuas are bigger than south polars but very close in appearance. Bill explains that the difference between them is a regional issue. South polar skuas – truly polar birds – are the primary predators, as long as there are no brown skuas. Far to the south in the Ross Sea area there are no brown skuas, and the south polar skuas there eat penguins. But where browns occur with south polars, the more powerful browns displace the south polars from their niche. South polars prefer penguin, but here at Palmer the browns monopolise the penguin colonies. They are the owners, and they won't tolerate south polars predating on their territories. Anywhere south polars are up against browns they become scavengers. But even while scavenging a dead penguin a brown will move them on. So the penguin meat seen in south polar regorge, and the feathers in their bolus, come almost exclusively from scavenging. In all his years here Bill has never seen a south polar skua kill a chick or pick up an egg.

Gentoos have taken over old Adélie colony sites on the ridges above the narrow loch where we lunch. While Donna and Chris census them, Bill takes Hugh and me tramping across rocks and through deep snow, real distances, to see moss beds and the old abandoned Adélie penguin colonies he excavated with palaeobiologist Steve Emslie in 1997. He shows us small sea creatures in a seam of soft sedimentary rock, stacked in the side of a narrow passage between granite cliffs, a lateral cut, with shale and softer muds, light cream and grey, splitting cleanly like tiles. Bill says they might be a type of squid with a partial hard shell. On Seymour Island, off the eastern coast of the Antarctic Peninsula, I've walked over bare yellow-brown hills, fossilised shark vertebrae and ammonites lying revealed like shells on the beach, exposed by the wind, late Cretaceous treasures. Bill reckons this Biscoe site would have been 15 metres beneath the glacier ice when Peter Hooper was geologising here in the 1950s. When, back in England, I tell Peter about the Biscoe fossils he is astonished and begs a photograph, any information. Hugh wants to investigate potential traces of organic pesticides in a moss sample, but the mosses remain intact, like the fossils. Biscoe is a SSSI site. Bill could bring me a thermos of water, but everything else stays where it is.

The new strait upsets Hugh. It is palpable evidence of climate change. The surface is crammed with ice floes and Antarctic action: crabeater seals ride the ice pieces in single possession; Weddell seals lie, silvery-fawn, on a snow bank. I'm secretly elated by this mini-version of sea ice. Palmer in summer is north of the main pack. Our ice cliffs calve off growlers, large lumps of floating ice. They don't bring forth icebergs. Large elegant bergs from ice shelves further south sometimes float by and park aesthetically in our bay, their bases grounding on the near-shore sea bed. Layers of broken sea ice come in with the wind: scummy bits, small pieces, clogging the bays then dispersing again. But from December the seas here are generally ice-free. Protected bays still provide a refuge for the ice, as do places like this, receiving the bountiful ice collapses from a fast-decaying ice face. Antarctic theatre. I've been missing it.

Back at our rock seats Donna has wine in a mesh bag chilling in the sea. We drink to Chris and to Hugh, who is determined to return next season. And to the best Australia Day I've ever spent. Donna sees three giant petrels, two with her bands. She's happy. She trots off to take photographs of the gentoos with her new camera. Everyone relaxes. Boondoggle time.

Four days ago, returning from Dream, we got soaked in bucking swell and salt spray. This evening the sea stays blessedly calm. Dinner is long past, and Debra Jo is raided again. Later, writing up my notes, the desk keeps moving. I've been out since breakfast: first in the big zodiac with Steffi for an hour and a half while she winched up grabfuls of mud, and now long hours with the Birders.

The *Gould* is 12 kilometres to the south-west. The Marine Projects Co-ordinator radios requesting permission to come in tomorrow at 7 a.m. The LTER cruise is almost over, and head of science Robin is 'wanting to get in as early as you guys were comfortable with'. Bob: 'The line handlers will be out at 7.30.'

I get up very early because I need to retrieve my gloves and boots left in the float coat room before the day's chaos. The *Gould* is already outside, waiting on the doorstep like the cat to be let in. The day fills with last-minute interviews, writing up Biscoe, the clatter and grind of cargo loading, people moving their stuff off the ship, departers getting theirs onto it and – US research ships are dry, and this trip has lasted three weeks – a loud, late, long party.

On Monday morning, grey and suddenly much colder, we the remainers gather in a loose cluster on the gravelly slope down to the pier. Friends from our small community are leaving; new people have come. Threads breaking, new threads to spin. It's the first shift in our local spider web. The departers say goodbye, walk up the gangplank, and gather in those well-remembered deck spaces, staring down at us. It's awkward, and special, and truly emotional. Heidi, newly returned, plays the trumpet, the clear sound of a single instrument. Doug, the Hazardous Waste Co-ordinator, suddenly sheds all his clothes and streaks to the ship as it pulls away, jumps off the pier into the water and swims among the ice pieces rattling in the wake, joined by Kristin and Jeff. Then he climbs up the iron ladder and walks barefoot and naked over the stones back to Bio without even appearing to shiver. And just standing there, in all my clothes, my hands are cold.

Most Saturdays, with House Mouse followed by All Hands Station Meeting, then weekend cocktails before dinner, often my turn at gash, I have a sort of half-day away from my desk. Last weekend I added the film in the lounge, a Station fixture, but shared spaces are becoming emptier as people take DVDs and play them in their rooms on their computers, or plug into them on exercise machines in the gym. Watching *Bridget Jones's Diary* in remote and affection-limiting Antarctica turned out not to be all that good an idea. Afterwards, as the sleet belted in against the side of Bio, I dreamed that my husband, Richard, was living in the other building, GWR, and my sister; and I wasn't including them in my life because I'd too much to do.

Now, as the ship disappears in drifting snow, I feel suddenly very tired. Ideally, I'd like to go away somewhere and cry; that would be best. But there's no privacy. I know it's the debt of Richard's cancer, diagnosed less than a year ago, three weeks after my father died; the months of operations and recovery; the worry and uncertainty whether I could or should leave him and take up this NSF place at Palmer; the work previously commissioned but not able to be done, so still to do; the decision finally taken to come when he got the all-clear; the intense last-minute actions involved in absence; the preparations needed.

Mending takes time. But very few people in this small, confined place are married or have left the needs and ties of family behind. I'm the oldest here. Today, I think, by a lifetime.

18

The inner circle

Tuesday 29 January to Saturday 2 February

Giant petrels have become land predators and are eating Adélie chicks out of the colonies. On Torgersen, Colony 1.0 has been wiped out; 18 has gone, although chicks could have moved across to 19, or been killed some other way. The Birders measure according to their protocols. They don't know how many chicks were taken: Bill guesses perhaps 22 from Colony 1.0, but that isn't a data point. I would like, as always, to track the detail. That isn't the science. How do they know? Bill attempts to help. He says he has seen giant petrels soaring over Torgersen and presumed they must be preying on chicks. You can assume that the skuas would only take one chick a day, so all the rest must have been taken by giant petrels. They can just walk in, remove chicks and eat them until nothing is left. 'There are 450 local pairs, 900 birds. That's a lot.' Bill does not want to speculate whether the reduced number of adult Adélies contributed. 'It could be factors at sea we don't know about.' The data for all study sites only reveal the figures indicating failure. The final coup de grâce is not revealed by Adélie egg and chick counts. When I ask Donna, she says, simply, that giant petrels are predators. Bill must get used to it.

But there are pieces of measurable evidence. The brown-khaki excreta of giant petrel chicks indicate the myoglobin of penguin meat, and last year's telemetry data also showed the giant petrels ranging far, then – as Adélie chicks began fledging – staying local. This year the data are even more precise. As a historian I think that, when large-scale forces dislocate society, local brigands often do the killing.

Heidi and Brett, back from working on the *Gould*, are being inducted into current Birder needs. There are quantities of data to enter, freezer loads of diet samples, poop samples and dead birds to be sorted and dealt with.

Next month the labs built for field biology lab work in 1971 will be gutted. There's a bronze plaque on a dim wall commemorating Dr Mary Alice McWhinnie, researcher into the biology, distribution and life cycles of Antarctic krill. In 1962 she was the first US woman scientist to take part in Antarctic field work. Twelve years later she was one of the first two women to winter-over at a US Antarctic base. Gossip says she was a nun. 'Not true,' says Maggie, 'the other one was the nun'. Now the labs, named for Dr McWhinnie, are being upgraded for today's technically advanced research – clean power and enough power, instruments run off computers, clean labs for DNA and RNA work. I'm seeing these rather scruffy, well-used spaces as they were: bits of Antarctic history, statements of once state-of-the art science. My hope is that a record has been made of what was, before it is no longer. The Senior Assistant Supervisor, Laboratory Operations, reckons she has more than enough to do organising the clearing of the labs in time and finding places to stack everything. Will the new labs be named for Dr McWhinnie? No one seems to know.

The Birder tent is awash with male/female rivalry. Territory has been demarcated with pink and blue tape along the floor, up walls and across the ceiling. Like Shackleton bringing a box of dressing-up clothes and make-up to Antarctica, the ever resourceful Donna arrives at Palmer equipped with what might be needed. The smeary lipsticks for Bob's dinner, lacy bras for his office door, cowboy kit and swirly skirts. Dinner last night was a birthday celebration for one of the young krill scientists, a brief burst of minor humiliation for her compared with Bob's. Impenetrable to me, until Maria, who has just arrived, explained American college rituals. After dinner the dressing-up supplies were raided by the women, who trooped along the walkway to a hen party in GWR. The wind began gusting hard. And I crept away and had an evening of delicious privacy, padding around with Bio to myself. Except that the men were somewhere, and no one knew quite where. None of this happened three years ago, but there were fewer females on Station. Currently there are nearly equal numbers.

Bill gives Heidi her appraisal: 'She's very impressive.' Brett is being taught the south polar skua work on Shortcut Island. Donna, approvingly: 'He has a well-ordered, highly retentive memory. It takes two seasons to train someone.' Field workers cannot be cannon fodder, as in some science groups. Birder recruits are like apprentices working with skilled practitioners, learning by doing.

Heidi comes up the hill to T5, and we talk on the porch in an elongated Antarctic summer sunset. Her initiation to the ice was 'grudge work' at McMurdo as a General Assistant – every day the roll of the dice, digging wires out of the frozen ground or scrubbing down spilt chemicals. Qualifications: are you trainable? Are you willing to learn? Immediately bitten, obsessed, plotting how to come to Antarctica again. Hearing about this mythical coveted place called Palmer, 'where you get to go out in boats and see wildlife. At McMurdo the Raytheon world was kept separate from the science world: them and us.' With a degree in biology and psychology, Heidi wanted to work in science. A winter General Assistant job got her to Palmer, where she heard Bill and Donna were hiring.

Heidi: 'You don't know the sense of *who* they are. Legendary.' She was hired by email to start work in October 2000. 'Bill was taking a chance. It's his life's work. But Bill and Donna make a decision to trust people. They are passionate people, therefore they impassion people. It's an honour to be trusted with their project.' Experienced Birder Matt came south to train her, with Chris. 'I was learning in the inner circle. It was invaluable. Matt could explain what Bill and Donna would look for: "This is the way that this is done. This is why." Bill pays attention to everything. He has the peripheral vision of being a biologist. So much of what he observes is instinct. He feels it.'

To me, Bill is constantly scanning. It's as if he is absorbing through his skin. He's a hunter–gatherer, adapting the hunter–gatherer skills of survival – visual memory, highly developed sense of smell, understanding of topography and the movements of animals – to the process of science, which brings in the meal. Except that back in Montana he does hunt, avidly.

The light is diminishing, and the porch is too cold. Amy skis past. 'It's hard as rock. I've never seen snow so hard. It's skiing on concrete.'

The Birders function as a team, but the team is welded and structured by the partnership of Bill and Donna. Bill: 'I think about the large-scale questions, what to work on, and try to ensure our funding. Together we develop a plan of how to do it. Donna executes it.' Donna is the undisputed organiser and logistics manager. She has a sense of the status of the season in her head, the day-to-day managing of the fieldwork. She pursues the data – worries them into place, makes certain that they are bedded down into their final form. Donna: 'There's no need for Bill and I to say anything because we understand what we are doing without words.' Bill: 'It's a wheel. The wobbles only come from the outside. It's why we can do so much. The partnership is an integral part of the successes during the last decade.' In public Donna defers to Bill – 'Dr Science' – and filters contact with him. Bill is inherently serious, introspective, complex. Dave Bresnahan: 'They are exact opposites in every way. Intellectually, and emotionally. But inseparable.'

Donna is a tough taskmaster. Her recall is impressive, methodology practical, notebooks models of neatness and clarity. For her, nine-figure band numbers are easy to remember. Errors are unacceptable; inaccuracy matters. 'Pick your fights', she says. Work spaces should be clean. Uncared-for gear is chucked out: 'I don't want science experiments in thermos flasks.' Her impatience threshold is low; in a rare moment of introspection she admits, 'I'm not that good with people.' But animals – animals are at ease around Donna. Her handling skills are extraordinary. Bill: 'One geep she was putting a transistor on fell asleep. It was snoring!'

The crucial issue, every season, is establishing continuity and consistency in the programme in the face of an ever-changing environment. Bill: If you want to work with me and be successful, follow the *S-013 Extreme Biology Palmer Station Handbook*. Don't do things on your own. People who have not followed the recipe book don't come back. It is the

standard by which we tie one year to the next, the way we establish continuity in the actual process of obtaining raw information, establish a
sense of familiarity to the study sites. We've never had a broken season.
We've always had someone here that can do the work.

And the crucial issue is ensuring full-time commitment. Significantly, almost all recent fieldworkers have come from the support side,
people who are already in Antarctica on contract. There's a pool of
people always on the look-out for a potential vacancy. Bill: Choosing
team members is very, very important. People have to learn fast – be
shown once, then do. If someone hesitates, they are out. Being able
to observe is critical. We're transferring information, but we are also
teaching how to get things from the animal itself. There are some very
subtle things to learn, feedback on how animals are reacting, how to
ensure that what you are doing has a minimal impact. And we must
train people who themselves can train others. The taught must be able
to teach. 'This place offers hazards: violent weather, sea ice, walking
on rocks, climbing cliffs, the response of animals. This is science in an
extremely remote environment. It's actually quite dangerous. Make a
mistake and help is far away or unavailable. You pay dearly. A number
of variables can kill you quite quickly if you don't know what you are
doing. We look for people who can assess risks, make decisions about
how to proceed – outdoor people, with a sense of confidence and self-
reliance. People who don't need an audience to be successful, who have
the ability to succeed in isolation. Intuition plays a big part in our
choice – references are useless. I've made some real blunders in the past.
Every year I get letters from lawyers, doctors, company executives, who
want to work for me. The message is, "I need to get away from it and
your programme is the therapy I need." I'm not going to serve as a
couch. They've come from a structured world. They have a high regard
for themselves. What's difficult here is everyone is struggling to maintain doing what they can, to deal with the isolation, the lack of day-
to-day diversions from home. You are struggling to maintain yourself.
You don't want to be faced with an individual who needs to drain you.
Don't come here with a huge ego.'

Very occasionally people apply wanting research experience. Bill is conscious of his lack of disciples, that he has not been involved in training PhD students, the next generation of scientists. He has no graduate students, no theses, under supervision. It's an issue, arising in part from the way he organises his work. He considers being a 'regular academic' to be constraining. He does not teach and no longer works in a university. 'I had to develop an unconventional way to carry out my research, to learn to support myself on soft money. My message is, you can make a living doing research outside the limits of the conventional academic world.' Fully decoupled from academic life, Palmer is in a real sense Bill's home institution. 'I realised a long time ago I wanted to work here. To devote my life to it. I saw how incredibly variable things were, not only within but between seasons. In a highly variable system long-term research is needed; the scale of the research has to accommodate to the scale of the system. I wanted to make a thorough study of one thing, not a little of a lot of things. I wanted to understand how this ecology operates. I needed to accumulate long-term information for things to make sense. That's the passion. That's truly where my energy goes. In order to do that properly I needed to immerse myself in the work. My focus is here. It's local. That's what I want to do.'

A long-term, people-intense database is being created. But, as Bill acknowledges, a research life is finite. Somebody must take his work on. 'They will have access to a thirty-year database. I would have killed for that when I began.'

Next morning the sea is the grey of whales, solid grey. Or the grey of the islands, as though their rocks had stained the water. Last night a gale got up, sweeping out the brash and small stuff, and now the only pieces of ice remaining are lumps of old glacier, each riding in the same axis, sculpted profiles above water, heavy bulk below, all pushed into position by the run of the sea like small boats tied to their moorings. White-capped, urgent waves in neat, angled rows move fast across the shallow bay, fetch shortened by the wind driving down off the glacier, kicking the water into a close swell. The ice glows thick white the way clouds glow against a storm-heavy sky. As I come down Bio's back stairs, a small square window frames the view, like a nineteenth-century New England sea painting.

Our local spaces, our intricate coasts, may well have histories of discovery by Yankee and British sealers, or Norwegian whalers, seeking profits. But the histories are not known. Now the ships spending time along these coasts carry passengers who harvest images and memories of beauty and wildness. Today tourism yields the returns, and the numbers increase every year. Tourism is a legitimate activity. The Antarctic Treaty ensures freedom of access to the continent but all tourist activities need to be conducted in line with the current Environmental Protocol to the Antarctic Treaty, which sets out a framework for the comprehensive protection of Antarctica, a protection that needs constant monitoring and reassessing.

An interchange in Russian cuts in over the two-way radio on my desk. Russia's old Cold War ice-breakers now earn tourist money in Antarctica, and staff communicate with linguistic impunity between the ships. There's a time slot, 7.30 each evening, when next day's schedules are negotiated between tour ships. No one wants to enter the peace of Paradise Bay and find someone else already in Paradise. No one wants to bathe in bubbling, warm volcanic water at Deception Island with unexpected neighbours.

Last night's northerly gale, bringing moist warm air with rain, is really shifting the ice off the glacier face. A massive ice fall pours ice crumble into the sea, and the impact wave surges powerfully out, across an inlet where the ocean is forcing through the new strait, to ricochet off the shore and move relentlessly back on itself.

A vast luxury vessel, on the second of two cruises in its first Antarctic season, is anchored by Janus Island, dominating the landscape like beachside skyscrapers. Another luxury liner is making six voyages to the peninsula this summer. The total number of scientists and support staff currently working at the three United States Antarctic Stations, and aboard the two US research ice-breakers, is around 2,200. The total number of scientists and support staff on all nations' stations in Antarctica this mid-summer is around 4,000. In eight brief trips into the

Antarctic Peninsula waters these two ships will bring many times that number.

Next season, Torgersen Adélies will be monitored for the impact of tourism, by attaching heart monitors to selected birds to measure heart-rate when approached by humans. The correct distance to stay away from an Adélie, according to US Antarctic conservation legislation, is 15 feet, unless listed by name on a permit, which everyone on Station is.

Do increases in heart-rate vary between the approach of a human and that of, for example, a giant petrel or an elephant seal? Bill deftly analyses the issue: critically, we do not know whether a change in heart-rate produces a corresponding effect on reproduction – that is, the numbers of final fledglings in the water. There may be evidence that tourism is affecting Adélies. Or no evidence. But we cannot measure tourism in relation to environmental variability: the presence or absence of sea ice, or snow, or the reality of the birds' foraging – what impact the sea environment is having. For a population to remain stable, adults have to be continually replacing themselves. But counting chick numbers, for example, cannot be used as evidence for or against the issue of the impact or otherwise of tourism, because the counting is not able to differentiate between the impact of tourism and that of the environment. Environmental data – the environmental history of a colony – affect the demography of birds. But environmental concerns are by definition long-term.

Twenty-four hours later an honourably battered small aluminium yacht is moored in Hero Inlet with six of the world's more experienced women sailors, French, British and Swiss. They have been the furthest south of any yacht this season. Yachts bring small numbers, their proportions complement the landscape. But they are private travellers, part of uncontrolled tourism, and there are rules. Yachts may not moor at our dock or take on water; personnel may not come ashore unless invited. Bob and Jeff are guests on board for dinner. In the night the women set sail, worried about a growler, drifting down onto the mooring lines. But next day they come back and are invited ashore for lunch.

At Palmer around forty people are perched on the edge of a glacier, supported by a complex and continuous web of logistics and planning. Yesterday, briefly pausing, was a world of 1,800 passengers enfolded in another complex web of logistics and planning. Our small Station is focused on researching our small, precise locality. The ship was world tourism, bringing schedules and mindsets that belong to the everywhere and anywhere of world tourism. Now six women – self-financing, unsupported, living by their expertise – brush past us.

The particular, subtly understood, give and take, trading of favours, insider culture of Antarctica. Co-operative, and competitive.

19

The sound of extinction

Sunday 3 February

Litchfield Island, at last, on a grey, moist, limpid day. Litchfield is an island from central casting but out of bounds, entry forbidden without an extra layer of permits. It has a skirt of open land, gently rising from wide rocky beaches, and a sense of real mountains – bold, high, rounded presences. The scale is impressive, mysterious, beckoning. But deceptive: Litchfield is only 2½ square kilometres, and no higher than 150 metres.

Rain begins, real drops falling on a pool of water stained with run-off from moulting elephant seals. Bill and I are walking up a shallow valley along the bed of a remnant stream, a trickle of water, the only approximation of running water in any of Palmer's islands. It's all that's left from the melt water of a glacier. The valley narrows into a high-sided canyon. Twenty years ago banks of ancient moss grew thigh-high in the stream bed. Litchfield was declared a Specially Protected Area in 1975, an outstanding example of the natural ecological system on the Antarctic Peninsula, to be entered only for compelling scientific purposes. But elephant seals came regardless, arriving in huge numbers, heaving their enormous girth and length over the moss, ripping it apart, reducing it to a squashed brief layer between the rocks, moss fibre felted with moulting hairs. Now the banks are all gone. Some moss still cascades on the canyon sides in soft-focus stillness, improbable confident green in the immensity of Antarctica's lack; gentle humps and billows, a depth of ancient growing hiding the sharp rocks beneath. Moss in motion.

Today we don't see many elephant seals. Four females are lying across each other, big brown eyes gazing at us, nostrils laced with white strings of mucus. An emaciated young male lies apart, inert, so thin his nose looks like a dog's muzzle, his eyes lacklustre and barely open. Bill briefly dreams of an illness that will affect nothing but elephant seals. It's the only animal I really do not like, he says sadly. They've done so much damage. But a few fist-sized mounds of moss have begun growing along the stream bed again. The elephant seal numbers have lessened on Litchfield. Bill speculates that the reason might be a reduction in krill, one of their food sources.

On a bank of snow I see what looks like an old glove. In my hand it's fawny, tawny, lustrous, like a piece of silk velvet. Tear it, says Bill. But I can't bear to. It's beautiful, the finest weaving, soft and light, draping exclusively. It is elephant seal moult. I turn it over, hoping not to find skin, and discover instead fine underhairs, dark-coloured. Then I do try tearing, gently, and it rips easily apart.

At the top of the valley there's a rocky amphitheatre where the glacier used to be. Here at the watershed water ran directly north and south, cutting through the steep, rounded hills. We speculate whether the hills could have been turned to these shapes on the lathe of the ocean's waves. The sea-level has been higher, in the past, and the raised skirts of the island are old beaches.

North of the watershed a small wide cove is filled with those elegant individual pieces of ice that demand you look at each — marvel at the shape, the texture, the colour, the ingenuity of design and construction, at the infinite variety of invention. The pieces jostle in the shallow water driven in by the storm of two nights ago, the heavy north winds that have brought this day of moist grey calm. Some of the ice pieces have been pushed high among the boulders of the rocky beach. Nearly thirty years ago Bill brought the gifted American photographer Eliot Porter to this bay and watched while Porter watched, and waited, and measured, and looked, and then — ready — took exactly one portrait of the ice, and the water. We find the bones of a seal. Five years ago Bill saw the seal, dead, on the shore. The following season its skeleton was picked bare by scavengers. Now the bones are slowly scattering. Near where we landed today a dead fur seal lay for two years, its body intact, as if it had just left life. No scavengers discovered it. Now, finally, it has begun to break up, Antarctic cold stretching out the process of decay.

The sand between the rocks is grey; the boulders are grey and black, except for one, a confident curve of smoothed pink granite. Where is the nearest pink granite? My mind veers across geologic ice ages, across immense glaciers lugging rocks in their icy grasp vast distances. Freedom to imagine, because Antarctica is a geologist's Christmas hamper, a continent of virgin questions, and no hope of enough answers.

South of the watershed the view is 360°, not blocked off into wedges. Nearby islands seem suddenly low, reduced, backyard stuff. Spring's blizzard-borne snow still lies thick on the south-facing slopes of hills. The summer is far advanced, but the drifts have not melted out. Beyond, there's a wide view of the Anvers coast and our promontory rising unexpectedly gracefully, its sturdy station buildings spread across the tip, the runs of pipes and webs of aerials, then around to the curving sweep of the glacier and the rough hills and dip where Base N, the first hut, was sited, and Old Palmer, the first American station, was built.

Stand here, on the generous uplands. Fifty years ago, a thousand pairs of Adélie penguins nested here, numbers estimated by the men rowing across from the British hut. Look down, to the ground, at the covering of small pebbles, each selected and carried in the beak of an Adélie penguin, sturdy pink feet clambering, slipping, hopping, gripping, climbing uphill, pebble held firmly, until it is laid, with careful intent, in a nest. Here is the reversal of nature's processes which move, over thousands of years, sharp-edged lumps of rock down slopes, fracturing and splitting, jamming and sticking, warming and freezing, until, rolled and thrumelled in the ocean, they are thrust back up on shore, sifted by the rasp and suck of tides. See the labour of penguins, the patient sieving and selection, the lifting and carrying. See the transference of seashore, to hillside.

The labours are irrevocably, poignantly, revealed. There are no sounds but the wash of the sea, the occasional calls of skuas. Every penguin is gone. The nests are abandoned. Listen to the silence. The silence of absence. The sound of failure.

Bill stands tall, still, on the carefully sorted pebbles. Standing where it should not be possible to stand, in the centre of a penguin colony, in the middle of summer.

This season, on Litchfield Island, only seven pairs of penguins managed to keep eggs until hatching. Eighteen days ago Bill counted four pairs of penguins in one colony, one pair in another. Five days ago seven penguins remained. This space was still theirs. Now they have gone. The sound of extinction is approaching. In two to three years, Bill says, Litchfield will be vacant.

How to pay sufficient respect to the passing of the business of penguin living?

In the end I look at a favoured nest site, well positioned on top of an outcrop, with a backing of vertical penguin-height stones, an extensive sea view towards the distant glacier face and a deep bed of pebbles. The ground is streaked with the evidence of functioning penguins, guano radiating out from the nest as it should be at all well-ordered nest sites. It is white, proof of fish in the diet, which Bill knows to be the case from diet sampling. The food is there in the ocean. But the penguins have left because they have no jobs to do. All this season's chicks that hatched – twelve of them – were taken by brown skuas. This year Litchfield will produce no Adélie penguin chicks. The destruction of the colonies has been total and complete.

Walking over the pebble plains, there's not much evidence of the struggle. A few bones lie scattered, some broken eggshells. I find a pair of feet attached to leg bones and a right angle of hip joints, like a macabre toy. But poking through the pebbles the guano is still there, pungent-smelling, thick liquid in its historic layers.

Brown skuas sit on nearby points of rock, quietly, wings folded. No angry cries, no diving, high-pitched alarm calls: they have no chicks to protect. The browns move in a little closer. I ask Bill what they are doing. We are the only living things here, he says, morosely. There's always a chance we might be dead.

To the outside eye the location appears to have real merits. A variety of small knolls for birds to build nests high above melting snow, an abundance of the right kind of pebbles, good access from the sea. The map drawn in 1957 at Base N, showing the locations of local bird colonies, marks six Adélie colonies on the south-east peninsula of Litchfield. When Bill first arrived at Palmer in 1975, those colonies were extinct; 884 breeding pairs of Adélies were nesting further to the west, but still in the shadow of the high central hills, still on the island's southern slopes.

With temperature change, with increased snow, the sites proved lethal. Storm systems tracking west to east between South America and the peninsula scoured snow from north-facing surfaces and dumped it in the lee. Penguin numbers declined rapidly. In October 2001 there were three or maybe four penguin colonies on Litchfield, down from ten when Bill began working here. The landscape model predicted disaster under certain weather conditions. Litchfield was expected to be the most vulnerable. This season the conditions occurred: harsh spring and early summer storms bringing deep snow, covering the island, hiding the contours of the Adélies' colonies, the rock ledges and high places where they built their nests. Continuing, relentless storms dealing repeated blows.

Here is climate change in action, Antarctica as a living experiment. Litchfield Island is a precisely located, clearly defined landscape, with just two key species, Adélies and brown skuas. Their relationship is straightforward; the numbers have been collected, and stacked in databases of sufficient length and acceptable accuracy. Contributing factors have been unpacked and understood, decline tracked over time. The hypothesis is clear, the outcome predicted. Data from Litchfield had already revealed that, whenever an Adélie colony dropped below a certain number, the chicks were vulnerable to predation by brown skuas. Litchfield has six brown skua pairs, the highest density in the study area. They are birds with long histories, many of them tracked since Parmelee began working at Palmer. One pair, for example, had a territory with 800 pairs of Adélies. As the population declined, they destroyed their meal table. This season Litchfield's Adélie penguins failed to hatch chicks and so failed to deliver brown skua meals. This season Litchfield's brown skuas fed until there were no Adélie eggs left, and no chicks. Nothing.

Bill needed proof. Now he has it.

The hypothesis predicted extinction for the Adélies, but there is an unexpected twist. Both predator and prey have failed to reproduce, for the first time, completely. Bill has never observed a disaster on this scale at any of his Palmer sites. Being Bill, he is cautious about why the Litchfield brown skuas failed. The Adélies arrived late, so possibly there was no food for the skuas when they needed to lay. Perhaps, if laid, their eggs chilled in snow and water. Or, if hatched, perhaps the adults were forced to fly across the sea to Torgersen to find Adélie chicks there, and absences from their nests were sufficient to tip the balance. Skuas prey on each others' chicks. The detail was not observed.

Climate change forces other change. Shifting weather patterns challenge the precisely balanced interconnectedness of living things, their dependence on established networks to find food, to reproduce – to survive. Litchfield is an indisputable case study of the impact.

But a new population is arriving. Southern fur seals, lithe, dark brown, big necks and shoulders thrust up above agile flippers, moving across the ground like top-heavy commas, their wriggling narrow bodies tapering to their tails. Sperm-shaped, testosterone-filled young males, taking over the crests of rocks, surveying the territory, twisting around and staring, challenging. Fur seals can move fast; they can climb; and every year more and more are appearing here on Litchfield. After being hunted almost to extinction, fur seals as a species have recovered, and their numbers are expanding rapidly. Six were counted on the Palmer islands in the first 1974 census. Now several thousand are swimming in from the north, trying out this available territory further south – evidence of the shifting ecological boundary sweeping past Palmer.

Staring out to sea, I suddenly see something unexpected. On the southern horizon, between two snow banks, is a silhouette. Then I realise that it's an Adélie penguin, just landed, in from the sea. It stands, head down, wings slightly curving outwards, shedding heat. And somehow the poignancy – to see something and think – How strange! What is that? – when it's only an Adélie.

But, today, 3 February, it is the only Adélie on Litchfield, where once there were thousands.

20

The south islands

Monday 4 to Wednesday 6 February

East of Cormorant Island two towers of brown rock rise abruptly out of the sea, each topped with a white icy pediment. They used to be joined to the ice cap by an ice bridge, which atrophied and finally collapsed in 1989, leaving remnant ice straddling both towers. Then the linking ice collapsed, revealing wonderful ice caves, perfect campsites, says Donna, until they fell in 1993–94. Now the remaining ice perches on each rock tower like two blocks of flats. The ice is sinking in on itself, and soon both pediments will be gone. A memory only, for those few who have been here long enough to have attended to the sequence.

The evidence for warming is transient. Proof by absence.

Landing on Cormorant involves a steep scramble up the rocks. At the top a snow bridge connects the rocks to a snow-drift. Slim, tall Heidi stops, amazed. She hasn't been here since December, when the snow reached above her head and stretched level across to the hills. Cormorant is the most distant and easterly of the five Adélie study sites, and the early season storms impacted heavily. The island's surface is ridge and drop – stripes of land in billows, high then low. Today the undulating rocky surfaces reveal the tell-tale pebbles of untenanted nests on the slopes. Adélies cling to the ridge tops, a scatter of mini-Mount Ararats. Only high cups of life survive like small arks, pathetic narrow lines with clusters of chicks and a few adults. Brown skuas perch nonchalantly, secure in their meals. The influence of landscape shows so clearly it hurts. Penguin numbers are so low Donna does the counting by herself, with no cross-checking.

We climb up to a biggish colony that appears to be functioning reasonably normally. But diminished numbers have split its cohesion, and a long untenanted stretch divides an occupied section at one end from a larger cluster at the other. Chicks stand in small crèches, a mix of hatching dates, fluff ball to half-feathers, the brown-grey down, dirty, bedraggled, pushing off in the inexorable and necessary beat to their growing. I wonder whether chicks will be scavenged out of the smaller section faster than the larger. But Donna says that the two groups won't be counted separately. It's not possible to subdivide and subdivide; there isn't time or energy. 'We will end up with a total number. Perhaps the chicks will scramble along to the bigger segment and crèche with the others.' She disappears behind a ridge to return with the news that a pair with two chicks, nesting alone, in a sub-sub-colony, seriously separated from a main colony, had gone. 'Bill has been watching them. He will be heart-broken.'

Brett and Heidi do what they hope are the last snow stake measurements of the season, working across rocks and old drifts checking snow-transect lines. In 1998 Bill described where he wanted to put the lines on the five study site islands, but they were not easy to set up. Some of the marker cairns with bamboos, 25 metres apart, were knocked over by elephant seals, some completely hidden by snow. A compass with correction for the southern hemisphere was necessary. There were problems with variability: rope sags and stretches when wet, and Brett and Heidi reckon measuring depth with any close accuracy is not possible. Protocols need tightening and refining. This season GPS is being used to establish snow-transect end-point data – exact latitude and longitude – on all five islands. Next season GPS accurate to within a few centimetres will be used. But working in Antarctica is difficult; practicalities continue to bump against data. Next time, says Brett, they will use a fibreglass rope that won't stretch. Snow depths are becoming increasingly significant, but they're a salutary reminder of the difficulties of setting up valid databases.

Elephant seals lie on an ex-mossbed, the ground rich brown with what looks unexpectedly, impossibly, like deep loam but which is instead ripped-up lumps of moss. Higher up, the island's single pair of brown skuas nest on a mossy grass platform littered with bits of penguin – bones, wings, carcasses. Heidi and Brett scoop up a grey-brown chick each, and Antarctic quiet is pierced by the guttural cries, the throaty warnings of the desperately wheeling, swooping parents. I hold the yellow field book high as an alternative target. Once, at the melt pool in Palmer's Back Yard where a club of south polar skuas congregate, a determined skua came at me from out of the sun, each run a calculated lower swoop, forcing me to my knees, until I'd got off its space. But as a group we are braver.

Heidi tries to tuck her chick's feet and wings together in a firm hold, to have two hands free to measure the bill, from the start a mini-version of the parents' effective weapon, strong, with a sharp downward-facing hook on the upper mandible for piercing skin and tearing at flesh. She positions a ruler into the wing feathers, while the chick squirms and the parents threaten. Donna repeats her regular warning – 'Don't get your face close' – and demonstrates how to tuck a chick across the lap using the elbow, freeing both hands. Chicks are weighed, the temporary green plastic leg band checked for loose sliding up and down the thin brown leg. I'm asked to hold a chick. That's OK. I don't have to measure its bill and wing feathers at the same time. But the angry parent standing on a rock half a metre away seems suddenly very close. We leave the skua platform together, and the gangly chicks scoot back into their nest territory.

Cormorant Island was named after its cormorants, the blue-eyed shags. The once vigorous colonies are mere shadows, the cream-coloured guano poured thickly over cliffs and ground the only proof of long and intense occupation. Nearly a thousand breeding pairs were nesting on the cliffs in January 1989, when the Argentinian ship *Bahia Paraiso* hit a rock, minutes after leaving Station, gushing fuel. The hull's grey smooth curve still breaks the surface in low tide, drowned helicopters still strapped in their hangars beneath. A constant faint sheen of floating diesel marks the site. Palmer's blue-eyed shags decreased dramatically. Bill looked for a precise cause: 'There are so many factors in ecology.' He observed that Cormorant Island's birds leave their nest sites and bathe daily in little embayments around the breeding colonies. The diesel became trapped in the embayments. Whales and seals swam away, but the blue-eyed shags didn't change their habits. They tried to preen the fuel off, and it got into their stomachs. Chicks, fed with regurgitated food, began dying two weeks after the spill. Hundreds of adults died, their plumage affected. 'It was their bathing activities that did them in.'

Now censusing is done on a little promontory, keeping disturbance to a minimum. The birds seem to be clinging to the very fringe of the cliffs – active nests, like chimneys on the outside edge of a roof, cream-coloured, made of mosses and seaweed woven through guano, chicks spilling over the tops. Counting only takes a few minutes. The parents move their long necks sinuously, grunting warnings. The chicks squawk, then grunt, copying the adults. At least the shags have had a normal season, with breeding schedules generally on time.

Three years ago, when I was here, the weather was so dry that walking on the guano felt like soft, gently yielding putty. Now, don't tread in it, Donna warns us, you'll go in up to your neck! A large cormorant chick walks towards us on its large inward-turning feet and determinedly nibbles my boots, works upwards, tugs at their laces and finds bits of my jacket to pull at. But none of it is food for a hungry chick.

Blue-eyed shags are in decline on the entire Antarctic Peninsula except for the deep south. Bill's hypothesis is that the decline could relate to poor survival of chicks after they leave the nest. If that's the case, the cause is probably their winter food supplies. There does appear to be a very real decline in the availability of Antarctic silverfish, *Pleurogramma antarcticum*, one of their most accessible prey items. Bill: 'We have a record of their diets right up until the 1990s. There should be a change evident in the species of fish they eat, and also the size class.' Evidence could be gained from sifting through the middens, unpacking boluses, except that that would require non-existent time.

Work in the south islands needs a calm sea, which the day has given us. We cross easily to Christine, the fifth of the Birder study sites, where the Adélies have suffered less than on Cormorant. The chicks are in the crèche phase, still mostly downy, only just getting into uniform, as Donna calls it – beginning to show their fledging feathers. Clumps are still hanging, obscuring the new white breasts, the shiny blue-back backs, the white curve around chins and necks that will distinguish them until their first moult as this season's fledglings. But chicks are beginning to wander. The runts, the chicks that are not making it – the ones that have not managed to get enough food, and the late chicks – are holding on to their colonies, standing by their nest sites, because that is where they expect to get fed. The parents are all out foraging. But desperate chicks will tempt fate, try going where their parents go, maybe in a food chase, down to the beach. Donna says chicks will follow each other, so a kind of fluid movement begins, with changes during the day. Older and more robust chicks can run faster and are less vulnerable, and they make the first moves together. They don't attempt to keep the others out; it's just that younger ones have less mobility. It's the crèche phase, but mobile. A population of thousands of chicks sort of behaving in the same way swamps predators. But little groups of twenty or thirty are instant targets. It's no problem for a predator. At Colony 8, still a reasonable size, the reality is stark: a swath of empty nest space cuts the ridge into two unequal, vulnerable sections, just like the big colony on Cormorant.

Adélie indicators are taken, the two chicks of the resident brown skua pair are measured, snow transects got through. Donna's fast pace has, if possible, got faster. The sea is so calm, the light so clear despite low-level cloud cover, that she has decided to achieve a census of kelp gull chicks in the south islands. So we cruise around the perimeters of Hermit, Laggard and Limitrophe, looking for sightings of kelp gull chicks on the cliffs above. It's a remote chance, because this year there are almost no chicks, but we do see several running up the cliffs, like mobile rocks, well camouflaged. The Birders do just enough kelp gull work to tick the data over. Disturbance is kept to a minimum, a very low-impact approach. Bill says it is monitoring, a labour of love, keeping alive a long-term data set started in the 1970s. There are never enough people to do all the work. But by doing things consistently a very good idea can be got of whether the kelp gulls are doing well or not. He reckons the adage that 20 per cent of the work gets 80 per cent of the data applies to the kelps.

Today is serious teach-and-remember time. Brett has this one opportunity to be shown landings and possible drop-off points, learn how the swells work in certain conditions, the hazards of hidden rocks. Donna's clear voice in show-and-tell.

Giant petrels sit along the cliffs, single heads poking up, equidistant. Antarctic terns flit around our zodiac, and Wilson's storm petrels dart as though hunting for insects – except there aren't any. A few yards away a leopard seal thrashes a large fish in its mouth, and fawny brown crabeaters swim, then haul out on floes. A Weddell seal, speckled-calm, lies on an island snowdrift. We boat through delicious pieces of improbable ice, clicking and fizzing, then turn towards the glacier edge and boat along close, checking Kristie Cove for gulls. A narrow waterfall of blue iceberg water tumbles noisily out of a crevasse above our heads, falling into an ice bowl and running back into the glacier to emerge somewhere hidden and gush into the sea. This is the drink of the gods, this centuries-old fall of clarity: it seems, somehow, appallingly wasteful.

Back on Station the weather reverts to rain and sleet, bashing all night against the wall by my head. I worry about Steffi, sleeping out in the Back Yard in a borrowed tent. Next day, wild and woolly, lumps keep crashing off the glacier face. Persistent slushy rain and the warm northerly air are destabilising the ice. Fieldwork is cancelled, and the Birders chip away in the tent at data entries, organise notebooks, plan next winter's two GLOBEC cruises. Bill is proposal-writing, the regular repetitive effort of securing funding for future research. Places to think and write are at a premium at Palmer. Bob: 'A private room to myself, some place to go where someone isn't looking at you, is critical. Essential, as Station leader.' Last week, talking to Bill in T5, one person after another discovered a need to come into this rarely visited building. I lost count. T5's minder found jobs to be done near my desk, even one that required standing on it. No one has privacy. It's one of the tensions of being in Antarctica. This huge unoccupied space, and little humans rubbing along in public scrutiny, and gossip.

I've been researching the spring and early summer at Palmer, the crucial days and weeks of Adélie hardship. Heidi and Brett show me their videos, a visual record to add to the evidence: data, impressions, interviews, written and spoken. But Bill and Donna require from these months only figures on their data sheets. The present and future drive them as scientists: the immediate past is just that – the past.

On Wednesday, at lunch, Brett and Heidi report more colonies on Torgersen with no Adélies left. A giant petrel has now been seen in the act of killing its lunch: 'Everyone is munching the chicks.' Bill is worrying that the transmitters currently fitted on two foraging Adélies may be affecting their feeding and wants to remove them. He watches his computer, monitoring the logger for signals that would indicate they are home. But despite searching, despite occasionally getting a signal, the two penguins can't be found.

I sit at my grey metal desk in my grey building wrestling with the complexity of what I'm trying to do. Outside the small windows with their skimp of tie-dyed curtains, a grey mist obscures the glacier face. The wind has dropped, but the sea is surging, high and energetic, and the sky delivers snow, rain and sleet, this season's trio. The air feels damp and warm. Warm today means a surface temperature of $1\,^{\circ}C$. We become tuned to the smallest gradations in temperature. Plus one matters because solid water becomes liquid. Ice melts.

The weather dominates. Its exits and entrances, its moods, its rages. We puppets jerk along on the end of its strings. The weather is the major player in our drama: it drives the existences of everything living here, from the newest chicks struggling to survive to the most experienced scientists struggling to achieve their schedules. I know that weather dominates in Antarctica. 'The A Factor', as Martin our voyage leader called it on my first trip south, with the Australians. All schedules, all plans and objectives, he kept reminding us, were subservient to the weather.

Describing the weather has become integral to my account. It defines and dictates our actions. Everything that has happened, from before the arrival of the first penguins, has been driven by the daily reality. The massive repetitive storms bringing heavy snow last spring and early summer wrote the pain the penguins suffered. Yesterday, showing their videos, Heidi and Brett searched for words to describe the impact. People I interview employ extremes. But now these recurring weeks of wetting rain and high winds, these repeating cyclones spinning in – this almost total lack of regular summer – are severely affecting the birds.

The seabird work at Palmer is lock-stepped to the weather. All field-work is done away from base, involving journeys by water, so is by definition vulnerable. But this season the weather is dictating what fieldwork is done, as well as the timing. The field trip to Cormorant and Christine was a poignant reminder of a normal day, in a normal summer; like last summer according to everyone's descriptions, or the summer I was here three years ago. But Monday's fine weather just forced an intense rush to achieve as much as possible while it lasted.

And there are additional layers to weather's domination. An arbi-trariness mixed with an exhausting relentlessness, an unnaturalness. Behind the work, the thinking, is the driver of the changing climate here on the Antarctic Peninsula. The way the changes are impacting on the seabirds, in particular the Adélie penguins, is where the relation-ship really binds.

Antarctica is original planet. It's what draws me back, the noise stripped away, the challenge to see, think, feel, in this uncluttered place. But it is Antarctica's starkness, this freedom from the complex-ity of much of the rest of the world, that gives it crucial advantages for scientists, as a place to study climate change. No cities, no agricul-tural practices, no highways, or change of land use. Here the physical environment is not relegated and regulated. The ecological networks, the food chains, are relatively straightforward, comparatively simple systems far from the confusing signals of most of the rest of the world. The requirement on every living thing continually to negotiate tem-porary occupancy, to manage the complex interplay of climate and place, is palpable. Living things flourish where they can, while they can. Salutary reminders for us humans, cocooned by urban living, lulled into assuming we can somehow ignore, or forget, the changeability and vulnerability of the thin layer of planet we use, the tiny, damp, curved space we happen to occupy at a pleasantly warm moment. Here on the Antarctic Peninsula impacts of warming can be tracked. It's a clear, stripped-down preview of what could occur elsewhere. It's an unpacking of the ways climate change can reveal itself. It's a prologue to the way climate change can happen.

At Palmer, this ferocious summer, we do not know the mechanisms delivering this weather, or how the weather relates to the peninsula's warming.

But I can document what it means to be here.

21

The year of reckoning

Thursday 7 February

The rain beats down. Proper rain. Water runs across the ground. Bill agrees with me: it's temperate rain.

The heavy snow earlier in the season had a fundamental impact on nesting Adélies and their eggs. Now these storms, with their soaking rain and strong winds, are affecting the late Adélie chicks – late because the parents bred late. Adults have water-repellent feathers with air pockets as insulation, plus their layer of blubber. Chicks need to build both thermal walls, blubber and adult feathers, using large amounts of calories. Cold doesn't affect the chicks. It's the rain, soaking their down, forcing them to shiver, using up vital calories in an attempt to keep warm. Bill: 'The rain will wipe out the weaklings, the runts. Soaked through, they shiver to death, too weak to respond. They can't fight it. Little bodies sprawled on their stomachs on the wet ground, barely lifting their heads. Immobile bundles, meals in waiting. The low numbers this season have already made them vulnerable. Now they are easy pickings for skuas and giant petrels. They just stand there, passive, being pecked to death. The older chicks can crèche for communal warmth. The little chicks are always on the outside.'

And Bill returns again to the way it was. Torgersen, the vibrant feeling of functioning colonies, the energy. As I saw them three years ago, a haze of down and pink guano dust, a cloud of excess, evidence of effort and success. This year there were only 2000 pairs of Adélies nesting on Torgersen. On 23 January the protocol chick counts for all colonies except 14 and 16 gave a total of 681 chicks. The production per pair looks like being less than 0.5. Three years ago it was 1.5.

Look at these colonies now, says Bill, and you can see the ones that are doing poorly. The dismal success of some of them. Small numbers enables you to see the individual. You don't have the large numbers to cloud your vision. 'What you are doing is feeling sympathy. Concern, pure emotion. If this had happened twenty years ago, I can almost guarantee it would have been an observation. Brain and heart separated. From day 1, young investigators don't have the same perspective. But as you get older, you get a perspective, a compassion for the animals with which you are working.'

I say, I think it's because the numbers are so sparse. My overwhelming desire when I first saw the colonies on Torgersen this time was to track each nest site, to see how each was getting on. I think it's down to caring about each penguin. It's no longer big numbers, but 'each and every one'. As in the hymn we sang as children. 'God cares for ...'

And Bill says: Exactly right.

We are in the empty Field Room. The gale-force winds gust rain horizontally against the window. In the usual search for somewhere to listen to Bill's download, to interview him uninterrupted, we've found two chairs and sit in peace under wet field gear drying on overhead lines. Birders change their clothes in the Field Room and prise guano out of the soles of boots. Penguin smells must be corralled and not penetrate Bio.

Down by the rocks the zodiacs shift and fidget, each harnessed either side by ropes clipped to an overhead wire cable, and a bow-line to shore. Bill nods towards the four big zodiacs, jogging and edgy, half a ton in weight, 20 feet long, each a miniature oceanographic platform carrying expensive electronic equipment. The krill boat tows a fish-finder, a transceiver attached beneath a float carved with a sea monster's head, a mini Maori war-canoe prow, telling researchers where, how deep, how dense krill swarms are, so that they can do a trawl. The sampling regimes here at Palmer for krill and phytoplankton are set-piece activities, sampling the same areas a certain number of hours and number of times a week, repetitive. Bill: Regimes like this can miss providing unique insights into the system. For me, going out into the field, all the

time, has been vital. I am here, I can make decisions on the spot, adjustments. If there are inconsistencies I can recognise them. The advantages
of flexibility have been clear. I don't know how else I would have been
learning about the system. The natural variability in the environment
comes to me.

It's one of the ironies I'm beginning to understand. Climate change
science needs data collected to the same protocols over long time periods.
Dogged persistent repetition is essential, but it isn't all. Interpreting,
intuition, fast footwork, is also part; questioning the new, the absent,
the unexpected. It's a complex mesh. I ask Bill about a theory I'd heard
as a possible contributor to Adélie decline. Bill's reply has typical apho-
ristic brevity. 'Outsiders play with ideas. Insiders deal with data.' Then,
elaborating: 'Ideas are embedded in anomalies. You make an observation based on a pool of prior experience. This becomes the basis of a
proposal for funding to do the research. This is part of working, and of
science.'

Current work involves young skua and giant petrel chicks. Bill: The
decision not to go out is made by the breeding chronology of the birds,
not by this weather. The rain means they are vulnerable to disturbance.
Potentially they risk their down getting soaking wet. We don't know if
in any way our activities prejudice their survival, but it's a presupposi-
tion. So it has become a personal rule: don't work under any wet conditions that could compromise animals and data. There are two aspects
– the moral issue, and the effect on our data, our end-product. Everything is a matrix. At the bottom is the question, how to create the least
environmental impact to get the best numbers. 'I'm always analysing
this. Almost every decision is based on this. Some calls are obvious,
some more critical. Birds really are being impacted by the wet weather.
So we have to decide which procedures to do – how to get something
accomplished by minimising the impact.'

The last two days have mostly been lost to fieldwork, but used for
entering data. 'We can recuperate. It's doable. But let's say a week's
storms – that would involve major consequences.' Data are sacrosanct.
The most critical component of the seabird ecology fieldwork is to leave

Station with the season's data in the databases, backed up and sent to the States. 'This crazy year, eight months in the field, has produced an incredible amount.'

But the new Adélie satellite monitoring programme has suffered. The weather has made heavy inroads on potential data. Bill has decided to limit each transmitter package to between two and five foraging trips per bird – it's all he feels he should take a risk on. 'There are lots of things playing against Adélies this season. They are clearly under tremendous stress. What I'm worrying about is that I'm very close to being obsessed with the consequences of putting satellite tags on these penguins. That's got me doing a 3 a.m. wake-up. It's the lack of insight about what is happening that is distressing. We know our methods don't harm the birds we are working with. But are they artificially affecting them? If we kill a parent, we are killing their chicks.'

Three of the tagged birds have disappeared, totally. 'It could be instrument failure. The tags are still reporting, at extremely odd intervals. Perhaps seals have chucked the carcasses but they are lying on floes. I've been living with this. One bird went into the sea after release, where he disappeared for the entire next day. Did a seal kill this penguin because it was more vulnerable? It was carrying a presence/absence antenna, so yesterday we went to Humble and carefully scanned the outlying islands with a portable antenna and picked up a very weak signal. Then the battery ran out. I was blind again. I went out later and picked up a weak signal again. It gives me hope. I can't tell if it's alive, but it is out of the water, somewhere.'

Last night Bill sat in front of his computer watching the screen scroll. 'At 9.37 two blips. My receiver was detecting a signal. I waited for twenty minutes, forced it to receive again. Then, all of a sudden, two of the three tagged birds show up.' Data revealed that after leaving Humble both birds surfaced locally: one at Norsel, one at Litchfield. 'For all we know that is normal behaviour. Or it's the need to recover. In the overall scheme of things, what's the big issue?'

Perhaps, I offer, given the stress of what has been happening to the Adélies he is focusing his worries on individuals.

Bill: 'I've come to know these birds so well. One of these penguins, it could be twenty-two years old. That's the equivalent of my age. It's successful. To come through this season it's very successful. It's an animal that's managed to be reproductive. And here we go and scum it up by putting an instrument on it. What right have we to be doing this? I'm not moralising. But I know the kind of investment these birds are making in their chicks. These are animals that are mature. They've done everything right to come through a ferocious season. And here I am, tweaking. I've never felt this type of tension. I can see the tension these animals are under. I'm totally consumed by this. And we keep on doing it. Every time – should I put another instrument out? It's a tremendous conflict. My emotional side is clashing with my scientific. This is the worst year for this tension, this conflict. Science pushing the emotions. The totality worries me. Whether they are going to come back. Checking the penguins. One bird I'm after, a Humble female, is on its third foraging trip, and it still hasn't come back. Everything else that seemed extremely important is now in the background. The complicating factor is that quantum leaps happen in anomalous years. Last year was blissful. Tons of krill. Sunshine. Elephant seals down. Predators down. But this year. This is the year of reckoning.'

We look out at the sea. This is a good place to talk, Bill comments.

The water is an extraordinary colour, like a bruise, a khaki – almost yellow-green inside the grey. It's full of phytoplankton. Each phytoplankton has a lifespan of, on average, five days, Maria has explained to me; yet phytoplankton produce half of our atmosphere's oxygen. Blooms happen all the time. But Bill says this is particularly strong. If we went down by the rocks and filled a beaker, the water would be so turbid we wouldn't be able to see through it.

I've seen phytoplankton released from collapsing floes in early summer, yellow-brown stains on the underside of the ice leaking out into the water. But here, coming to my Palmer door, is a great burgeoning of the ocean's plants, a generosity of growing, a massive pulse of reproducing. The engine driving the food chain fills the sea. The natural variability in the environment coming to us.

We step out of the dive locker at 5.30. There's a smell of wood smoke from the galley stove. An evocative resonance with other places where it's cold and blustery and wet, other west coasts.

In the lab the laptop is on, logging data. The blips of the missing tagged female Adélie appear, arrived on Humble. Straight after dinner Bill and Donna go out, capture her and remove the black box from her back.

22

Crunches and crunched

Saturday 9 February

Yesterday the track to T5 was skiddy with small rocks prised by running water out of holes they'd been frozen into for years. Now thick snow drives almost horizontally, and each piece of dislodged rock has a bulge of white. The glacier face has disappeared. Another day of severe weather.

In the tent Donna shows me her backpack, a tightly wedged cave of goodies and back-ups. She knows everything in it: hand-warmers, spare gloves – two pairs of inners and her gloves for animal work – home-made neck-warmers velcro-ing together around the chin or over ears, waterproof trousers cut off at the knees for a dry bottom if waves come into the boat. There are sharpened pencils, a nest tag needing to be returned, bits of food, throat lozenges. A fur seal whisker like a piece of fine plastic, thinning evenly. Donna is a survivor. Even now she's handing out a vitamin C lozenge to everyone, and a French gherkin, jars scavenged from a recent tour ship.

Every twenty minutes Bill wakes his laptop, hoping to find the still missing tagged male Adélie. Co-ordinates from the satellite have given a probable location for the faint signal, in the distant Joubin Islands. 'I think he's staging his return. He'll be in soon. I'll retrieve the transmitter if he shows up, whatever the weather – Humble is always achievable.'

Brett comes up to T5 to tell me about his work at the start of the season. The snow is 30 centimetres high outside the blue door as he arrives. An hour later it has disappeared. The air temperature has to go just above freezing here, Brett says, having been just below, and the snow evaporates.

After lunch Bill offers another download in the field room. 'We are in a terrible crunch right now.' All work relating to Adélie diet and foraging studies is behind. The diet sampling I saw on 21 January showed a sharp move from all-krill to a mix of krill and fish. Ten days later the next samples, taken on Humble, showed more fish than krill. Now a complete round of diet sampling has been missed. Bill: 'We haven't been able to get out since. If the birds are all eating krill again, is it because krill are available again? What did they do in the meantime? When did they change? There are so many questions. The loss means we do not know what the birds were eating at that stage of their lives. We can't recapture it. Every year we try to sample close to the same date – then here is a real blank. A zero. Every season we fail to get some numbers. We can try to achieve them opportunistically, but it is important to accept the loss. Each year what we miss is different. We do try to optimise the continuity of databases within seasons and between years. The absence of data is important.'

The most critical crunch is the weighing of departing Adélie fledglings as they leave land for the first time and enter the sea. Chicks fledge according to their age. The weighing is done on Humble's main beach and two subsidiary beaches. The average date for the start is 2 February, but this season fledging is occurring the latest ever recorded, and the first weren't weighed until yesterday, Friday the 8th, on Humble's main beach, during a break in the weather: fourteen birds, compared with the usual hundreds. Fledglings may have been there a few days earlier, but no one could get out to look. From now on, every two days, Humble fledglings will be caught and weighed in the morning. Weighing is priority. The only thing that can stop it will be winds in excess of 30 knots; wet weather is irrelevant. But Bill will push the limits to get the weights.

This year, the year of the harsh natural experiment, fledging weights are critical for Bill's climate change hypothesis. 'This season will allow us to sort out the extent to which the *timing* of the Adélies' breeding – influenced by the landscape – affects their weight at the moment of departure, newly adult birds, fifty-four days after hatching.' The krill population is seasonal for a major part of the chicks' feeding, and in heavy snow years when Adélies breed late and chicks hatch late, there is a greater chance of parents having to feed chicks when krill numbers are decreasing. But the timing of the presence of food is only one factor. Availability and quality are also significant. 'There are so many variables at play here.' Were the fish in the Adélie diet samples this season a blip, or did they represent a less dense distribution of krill than, for example, last year? The loss of one week's series of diet samples is frustrating. 'Our study is as close as any to sorting some of these things out. We have fifteen years' data on foraging trip durations, and twenty years' data on Adélie diet. In essence, the fledging results will summarise the whole season.'

Fledging weights are the last Adélie job. After this, says Bill, with evident relief, 'Adélies are done.'

Initiating all the current seabird strands and keeping them flowing and tracked is complex. Everything depends on access – getting out to the islands, and availability – sufficient labour for the work. Plus now, of course, the work is extending: back, to the winter that led into the summer; forward into the autumn that becomes the winter; out, to potential winter feeding sites; down, with Adélies, into the sea; up, with giant petrels and skuas, into the skies. All involving extra time, new work strands, data needing to be captured, followed and processed.

All seabird data are recorded in personal field notebooks, organised individually. From here they are transferred to a shared large data sheet, photocopied so everyone can check, then on into computers in a methodical, concentrated, accessible form, kept secure. It's a triple process, each point of transfer a moment for checking, for interrogating personal data and picking up mistakes. Both transfers are a form of proofing.

Bill: 'To many people on Station we only do fieldwork. It's visible.' Donna: 'People think when we are not out in the field we are not working. They think time not out is time off. They don't see the data management and administration we do in our office, the tent.' The fourth activity, lab work, has its own complexity. Bill: 'In the lab we are an "open season" for criticism and comment. If I was here at Palmer and doing something with test tubes in Lab 2, I'd be a scientist doing responsible scientific work. People have a difficult time understanding that there is another kind of science. The kind that you don't test in a lab. We don't work in a little cubby at Palmer. Our lab is outside. Sometimes we contemplate going to the zodiac in four lab coats. But now our lab has decided to kick a little weather out – and we can't go into it.'

To Bill and Donna each day's activities are self-evident. Bill can't understand why anyone would find his recipe book – the *S-013 Extreme Biology Palmer Station Handbook* – anything other than clear. It's a mystery to him if anyone has difficulties. I've read it three times, and I'm getting there. But so much has to be understood and I grab chances to ask. How many Adélie penguins, for example, are needed to maintain a brown skua territory?

Bill explains with educative clarity. 'A brown skua does not measure the number of Adélies under its control. It is defending what is defensible. It's a compromise. A large territory is too much to defend. Too small, and there's not enough feeding. Brown skuas take what they need.' Three summers ago on Torgersen I watched an Adélie run out of a packed colony chasing a brown skua carrying off a chick, and – in the open – fight it and secure the chick. In a depleted colony an adult chasing like this would just provide added opportunities for predation. Bill: 'A large Adélie population can absorb the loss. But the skua take remains at the same level whatever the population; so as Adélie numbers shrink the take becomes a higher and higher percentage of the total.'

Some procedures in the recipe book have been cancelled. From the late 1980s to the early 1990s, between 50 and 250 Adélie chicks per week were selected from the same general area and weighed. Bill: 'We established a growth curve and published the results. But the amount of labour was high, and I didn't feel comfortable with the level of disturbance the activity created.' Flipper-banding, that long-established technique to identify and track individual penguins, has also been cancelled. Bill argues that it definitely does harm, and that the evidence that banding affects results is overwhelming. At Palmer, from 1989 up to a thousand fledgling Adélies were selected each year from the Humble colonies and fitted with a metal band, each with an identifying number around the top of the left flipper, as part of the CCAMLR protocols. Only banded fledglings were weighed and measured at the point of departure, to evaluate their condition. Returning as adults, they became known-age birds, KABs, providing a database of life histories, as well as data on reproductive success. Bill made the decision to finish flipper-banding at Palmer in 1999. The CCAMLR protocols still include it, and other US researchers still do it. Bill: 'We have nearly 2,000 records of what banded birds have done over the last decade plus. The only thing we've lost – we don't know, can no longer track – are age-specific parameters. My philosophy is, if you are doing something to a bird – an instrument, a band – always ensure that you are looking at a population where you are not doing it.' From the beginning at Palmer Bill had a population with bands, and one without. 'The two populations constantly checked each other.'

Tomorrow, Sunday, the Birders intend getting up early for the second round of fledge weights, weigh the Humble giant petrel chicks then, weather permitting, census the Adélies on Cormorant and Christine which hasn't been done since Monday, the last day of fine weather. Now it's overdue.

Sometimes during a conversation Bill says I make him think about wider issues; or, the talk is helping him stand back and reflect on what he's been doing. And I get vital insights. But Donna says, sometimes Bill has had a brain lavage. He comes to bed silent, and doesn't speak again for days.

On cue Donna bustles into the dive locker. In spite of the weather there are diet samples to be done and finished before supper, because the Birders have put their names down for gash.

'All birds behaved a bit unruly,' Heidi reported later. Only one of the five penguins lavaged was female. One penguin had eaten only krill, one almost entirely fish, two a mix of fish and krill and the fifth fish, krill and hundreds of little amphipods. They hoped to achieve two objectives and see the tagged penguin but failed.

The work in the lab sorting the results of five penguin stomachs takes till 2 a.m.

23

Losing days

Sunday 10 to Friday 15 February

On Sunday afternoon the *Gould* arrives, racing its bigger more powerful rival *Nathaniel B. Palmer* to Palmer and winning by thirty seconds. The *Palmer* is here on a quick visit during a six-week marine geology cruise, a bit of R and R during a marathon stint at sea. 'Like cowboys out on the range back home in Wyoming coming into town', says Jeff. Whale watchers from the *Gould* haul themselves one by one to Bonaparte Point on the heavy trolley strung on pulleys across Hero Inlet. People on Station might expend that kind of energy working out on the rowing machines. But across the scientists go in the wind and rain, scrambling over the rocks in their red survival suits, looking inappropriately large. We aren't used to seeing humans on our immediate horizon – only elephant seals and skuas.

The satellite pictures have been showing a storm curling down towards us from the tip of South America. Yet another storm, except that this one is big. The weather has been poor to appalling, but there is a sense that this is an increase, more of the same but with a presumption of greater intention. Station members are invited on board the *Palmer*, moored in Arthur Harbour, and we climb into the zodiacs in a 40-knot wind and driving rain as the storm begins. Going out to dinner, Antarctic style. At the *Palmer* we scramble up the rope ladder and duck under a long metal pipe curiously placed to stun, except that it's part of the current science equipment and so has to be there. This is the Jumbo Piston Corer. It can fall like an arrow through the water, plunging into the ocean floor 1,500 metres down, then, exerting enormous pressure, fighting the suction, remove from the reluctant sea-bed a core 21 metres long of accumulated mud. Marine geologists on board are trying to discover aspects of the last 15 million years of the continent's history by analysing the layers of fine sediment hauled up in the cores.

At dinner in a cafeteria-style galley large enough to hold eighty-five scientists, I sit next to an Italian geologist who had been destined to follow his father and two brothers into the army, except he failed the exams. Be a doctor, said his father. But as a student he went on holiday, saw rocks and said, that's it. His family, he laughs, were appalled. Now he's working in the USA with one of America's leading marine geologists. At 2 a.m. tomorrow they leave for the edge of the continental shelf, 225 kilometres out in the ocean.

T5's knowledgeable long-term science technician has arrived on the *Gould* replacing Orion. Station gossip warned him about the intruder on his premises, but of course I vacate my grey desk in time. I have a new perching place amid a different kind of technology, blessedly quieter, and from today I sit on a stool at the high lab bench in the Dive Locker, my computer just fitting between a large wrench and a deep sink. Now I only need to walk twenty-four steps from the float coat room across the track and I'm here. The filtering coffee in the galley and the six dozen cookies baked every day of the week, laid out on the counter with a cake or two, are perilously close. But my time is limited. The Dive Locker is being taken over as a store for lab equipment while the labs are rebuilt.

Tonight is a sociable Station night: old friends in the bar, good conversation. Palmer will be crowded till the end of the month, with not a spare space. Steffi is happy to have Randy, the Marine Projects Co-ordinator on the *Gould*, on shore. They're marrying in the summer. But it's not great weather for the tent.

Monday the storm is fully into its stride, winds gusting to 55 knots, and hard, continuous, driving wet rain. Air temperature is 2.5°C rising to 3°C. Rain in these quantities is unheard of here. 19mm has fallen in the last twelve hours. There have been no equivalent twenty-four hours in February like it, in all Palmer's record-keeping.

The *Gould* is on a lee shore, and Captain Robert cancels all cargo handling. The ship was delayed twenty-four hours in Punta Arenas while lithium batteries were flown down from Santiago in a specially chartered cargo plane. The batteries are essential for science work on the current cruise, but military demand limited availability. Now the lost day means the ship is here during a storm and risks losing more hours.

Who absorbs the loss? 'Who eats it?' to quote Dave Bresnahan. There's a little stretch in the ship's schedule – some weather contingency time. But everything in Antarctica is time-sensitive. Ship schedules are planned up to a year ahead, scientists fine-tuning their lives to arrival and departure dates, support staff juggling contracts. The detailed logistics of cargo supply and handling, of fresh food consumption and deliveries, of science programmes lined up like aircraft on a runway – all are time-sensitive. Dave: It's a fairly complex system of dominoes. Everything, like everything else in Antarctica, is interconnected.

Typically, the lost time has to be absorbed by the participants on a cruise. Scientists working in Antarctica need to be flexible. They must arrive with plans B and C in their pockets. Almost certainly they will have to make adjustments to their hard-won ship time, modify aims, lose some opportunities. Politics will flourish like phytoplankton growths in summer, schedules will flip within hours, and wily players get more of what they want. This is the nature of doing science in Antarctica. But there are unwritten rules. Researchers must be careful not to go into other people's territory, use equipment being kept for other scientists, 'foot in the door stuff'. It's called 'doing whatever you want' and is considered not fair. Boundaries in Antarctica are closely watched, patches jealously guarded. And I get a sudden insight. The first time I came to Antarctica funded as a writer, I researched a second book as well as the one I'd been selected to write. I thought I was giving double value. But there were critics. I hadn't been selected to do *that* topic. And, at last, I understand.

At five to two, wind easing, we form a human chain moving food from ship to storage. 'Freshies time.' Everything is already a bit less than fresh after the journey south. Unloading complete, the All-Call sounds for line-handlers. Only support staff, covered by insurance as employees, are permitted to lift, disengage then release in sequence the heavy ropes holding the ship to bedrock and stays.

Twenty minutes' effort, and the *Gould* is gone. Fieldwork has shut down, and Birders work in the lab processing frozen winter diet samples taken during the winter GLOBEC cruise. Donna worries about the amount of lab work still to get through. Bill worries about the south polar skua chicks in their scrapes of nests – barely dents on the surface. Only three or four weeks old, scruffy, big tready feet and barely formed wing feathers, sharp little beaks and bright eyes – the rain must be soaking them. They could be wiped out.

All the Birders' south polar skua work is carried out on Shortcut Island. Labour-intensive and meticulous, it's part of long-term database building. Every five days approximately seventy-five numbered nest sites have to be checked for egg and chick counts, plus whichever known-age birds are making a first attempt to breed. Skuas are always on the prowl, ready to take each other's chicks opportunistically, so time spent in each skua's territory must be minimised. This miserable season south polar skuas bred late, so chicks hatched late. The work is desperately behind, and has to be fitted in somehow. But the harsh conditions are taking more concentration, increasing the risks of working. Shortcut is particularly difficult territory in snow, ice and rain.

At least, Bill says, the giant petrel chicks are old enough not to be affected by the storms. No longer brooded, they sit in their nests, compact and still, wings folded against the wind and the weather. A few have an adult sitting near, providing some body warmth.

On Tuesday we wake to white islands and five centimetres of snow. The third round of fledging weights is achieved, but only thirty fledglings are on the main Humble beach. There's evidence of more Adélie chick slaughter, further depressing Bill. Humble giant petrel chicks are weighed and measured. Brett and Heidi continue lugging the backpack containing the GPS over islands, establishing more snow-transect lines.

This evening everyone crowds into the lounge to watch a movie about

a West Virginian Elvis impersonator called Jeffco, a Station favourite
– relax and laugh time. Donna dresses up in hillbilly clothes. Unwarily,
I tip into my old world. I think the movie is well filmed. In Antarctica
I attempt to damp down the complexity of cultural baggage, limit dis-
traction. I've brought no music with me, no books. Yet of course I fail. I
try not to describe the physical world by analogy, to use metaphors from
our irrelevant cultures, but they keep jumping out. Curiosity spins me
into people's lives. I need to understand whatever bits about himself
Bill chooses to proffer. And of course I am a cultural artefact. I speak,
although I wasn't conscious of the details, with the vocabulary of where
I live in London, and the people I spend my time with. Which, I'm
told, is full of emphasis and enthusiastic exaggeration. When I go back
home, I hear the phrases all around me. I pronounce words according to
the same constraints. I say 'Weddell' (the sea, and the seal) in the way
it is pronounced in Britain, and not, as I'm informed I should, with
the accent on the last syllable. I don't pronounce the Neumayer Strait
as 'New-Mayer', or 'Add-ell-ee' for the penguin. None of these things
matters; they are culturally determined. But I'm becoming self-con-
scious. The season's stresses are beginning to insinuate cracks.

Then, under my mattress I find a book, an old, worn paperback. And
I break my self-imposed regime and begin reading Kafka's *Metamor-
phosis*. 'One morning Gregor Samsa awoke from uneasy dreams to find
himself in his bed transformed into a giant beetle ...'

At quarter to five on Wednesday morning the fire alarm rings harsh
and insistent, which means get out of bed, now, and muster in the
Boat Shed immediately, fully dressed because this may be all you will
have to wear. The roll is called. This is not a practice, as has happened
twice already, and staff trained as firefighters before they come south are
running between buildings in their specialist clothing, which hangs on
hooks by doors, ready to go. On the US Antarctic Program women are
as likely as men to be driving JCBs or doing marine tech work. And
now these 25-year-olds are firefighters. With intense relief, it's a false
alarm. T5 has a new alarm system, and it has been set off by a heater.

Jeff begins getting the zodiacs ready, jumping from one to the next in his float coat, pushing and scooping the snow off with a rubber dustpan (it floats, and doesn't damage the surface), then checking the pressure with the air hose (most boats gently seep, but ours, being black, also collect radiant heat). If there's a chance of his enemy, pack ice, he gets up in the night. The boats are harnessed, and they can't stand pressure from compacted ice or random bergy bits. Boats are central to Palmer science, and every morning by 8.30 a zodiac has to be ready, fitted with each science team's specific requirements, if and when they need it.

Boats are moved in and out of the water in a constant programme of maintenance, switched overnight. It's 'magic'. When he became boating co-ordinator, Jeff studied all the manuals he could find on his charges, and now his passion for zodiacs is catching. 'They are beautiful – serving the purpose of a boat, holding people – yet they don't depend on their shape to maintain their buoyancy and can float all ways, as long as air is in.' They're made of heavy-duty fabric embedded with rubber, similar to hot-air balloons. He shows me how the aluminium floor is in four parts, 'hinged' by tongue and groove, so it can flex.

But Birders' boats take the most time and effort. Bill and Donna prefer the 16-foot boats with 25 hp engines – easier to manoeuvre, they can be pushed off rocks, if necessary. Jeff: 'Birders can't do their job without being tough on boats. They work between stone and water, an uncontrolled area, at the whim of wind, tide and rocks. You become tuned to the maintenance aspect of their boats. After seven to eight years' use they wear out. The Birders don't receive the praise they deserve. This is a unique place but dangerous to work. The weather is fickle. It never makes up its mind, changing rapidly, and there are no weather forecasts. There's frozen water everywhere, random dangers, things floating all over it. Shallows everywhere. Precarious places to land, difficult terrain, complex topography. Bill and Donna have a knowledge and experience of this place greater than anyone. They take advantage of every opportunity. People use them as a gauge. I try to let them know that I have the utmost respect for their knowledge. I can't tell them whether to go out or not.'

Thursday is Valentine's Day, a friendly event at Palmer, with iced donuts for breakfast. Here everyone is everyone's valentine, no problem about whether anyone will remember us, because everyone remembers us. We all must have people and things we are missing, but the convention is generally not to talk. Maria tells me about the constraints on her ice time of two growing offspring and a husband with professional commitments. 'All my trips away have to be thought through very carefully. This time I'm here for a month.' We recognise each other's shared reality. She hasn't been at Palmer for two years, but conceptually she's moved things on. I tell her about the assumption among people we know in England that my husband would of course be coming with me on each Antarctic trip, and we laugh comfortably.

The people who arrived on the *Gould* are doing Boating 2, attempting to pull Jeff in full-immersion suit out of the water into the zodiac. I pass the method on to son Tom directing a film for the BBC in London, and Daniel Deronda pulls the drowning Myra out of the Thames in her soggy brown dress, rolling her into his light skiff in correct rescue-from-a-zodiac style. Here, with water around zero, Jeff instructs us that we can last some time in a float coat with a beaver tail, as long as we get into a huddle position to keep the body's core warm and don't try to swim. That's dangerous. The more you work, says Jeff, the less time you survive because it drains the body's heat.

Birders achieve the fourth fledge weights, finding only 30 to 40 chicks on Humble's main beach. Then, tackling the south polar skuas at Shortcut after an unprecedented weather-driven, eleven-day gap and a complete missed sequence, Donna slips, badly bruising her coccyx. Being Donna, it's not something that's talked about. Bill: We are beginning to look at cancelling things because of the weather. Perhaps the skua satellite work can't happen.

The fire alarm jangles at 1.30 p.m. on Friday: another false alarm. Mustering in the Boat Shed, I see a Birder zodiac upside down on the work-bench, with customised chafe guards glued on vulnerable places under the bow. Patched cones show where leopard seals have chewed. Jeff sews chew-preventers from heavy kevlar, except that he's run out and so uses rubber doormats.

The Birders are out on the long haul to Dream, taking Steffi, who has achieved today's boondoggle. Donna works in the tent resting her injured back. People love to be given a hard time on trips with us, she says. Then they can talk about it: it's 'their' Antarctic experience. We skirt around a complex subject. I need to go on field trips to observe the work, and yet I have the same responses as everyone. I know when I talk about Antarctica I'm likely to focus on the extraordinary privilege of being accepted or at least tolerated by animals, that prelapsarian encounter. The Birders' work is enabling. It allows access to the islands. It allows the touching of animals. The desire to be involved with them has no parallel with other science teams currently at Palmer.

In the labs, IT Hugh shows me the ceilings. 'Look up! Wires looping and hanging. It's all ad hoc, torture to deal with.' New networking equipment is being added to the buildings. Palmer, small, with high costs, has lagged behind the other two US stations in computing and networking. 'Science is all about data, and computers are now a critical tool on everyone's desk. But the equipment was sent down without the fixings. We're using wood chocks and plumbers' tape. The usual Antarctic making do, love/hate. Highly paid people doing mundane jobs.'

Young graduate scientists here spend most of their time in a lab. Handsome Michael stares down a microscope at the eggs of gravid krill, counting how many a female has produced, measuring what size they are. But, at least, 'they are not just anonymous eggs': he experienced the krill being hauled out of the sea during the LTER summer cruise and sorted through the catch – the familiar mix of physical work and technology. Lisa, expert cross-country skier, works in a small dark space processing phytoplankton samples, largely alone. There's no sense of the world just outside: chemicals and mildly radioactive substances, not islands, and birds. She tries to take one day off a week, white coat to ski gear. One morning Lisa passes me with shining eyes. 'I get to be a Birder for the day!' 'Where are you going?' 'I don't know. A bunch of islands I think. It will get to be a long day.' And afterwards, when I asked how she had got on, in almost reverent tones: 'I got to hold a penguin.'

An apparently calm and drizzly day turns, and the wind builds up. Donna listens to interchanges between Bill on Dream Island and Dave in the comms room. 'Bill is talking into his coat, like penguins talk into their wings. I know; I can hear the wind in the mike.' Later, walking into GWR, I hear Bill's disembodied voice coming out of a work jacket hanging in the lobby. A radio has been left on in a pocket. Bill is deliberately addressing the public ether. A film-maker has tracked him down and he's saying he's behind on his work, the weather is changing and he can't spend longer with him.

After dinner we go to the Carp Shop, the high open carpenters' space, scented by wood shavings and linseed oil, to tie-dye. Almost everyone did it at camp when they were six or eight, but they show me how. It's memory time, co-operative and sociable. A few evenings ago we made paper, producing satisfying indigo blue and newspaper-grey oblongs: thickish, bumpy, but undoubtedly paper. Now everyone brings spare white T-shirts to bunch and dip. I've travelled light, so I convert two veteran silk underlayers from grungy cream to psychedelic swirls and purple-green camouflage shadowing, instant ageing hippy.

Tomorrow, Saturday, I'm watching the fledglings being weighed. Bill is careful about the portrayal of his work, forbidding photographs of lavages, chases and nets. But he says I can describe any procedures I see. 'There is enough proof that weighing does not cause problems.'

24

The weight of a fledgling

Saturday 16 to Monday 18 February

A wide wedge of brash ice hems us in, slow-motion heaving reaching almost to the islands, clicking clacking clattering as the swell sucks in and out. Glowing greys and suffused greys and reduced whites, remnants of last winter's floes jumbled up with lumps of old iceberg and larger slices of ice face, chips and bits filling the spaces between, like thick vegetable soup. I listen to the delicate hissing as historic air trapped inside melting ice escapes. I am ice-starved. Milky ice pebbles and shiny water-worn ice boulders lie tossed up high on the rocks, the end of their journeying.

All night the wind brought the brash in, a silent passenger. Winds and currents are always moving rafts of sea ice around, bringing them north, then south, into bays, up against the land, disbursing them back out. This isn't the constant pack ice, whose seasonal boundary is currently about fifty kilometres to the south, waiting to grow and spread with approaching winter. It is residue, accumulation, the fragmented edge, a kind of unpredictable Antarctic sweeping, a reminder of winters past, winters to come.

But out on the Humble beaches this season's maximum number of Adélie fledglings could be congregating, ready to leave the colonies. Today could be the critical 'peak fledging'. On the other hand, this difficult year there mightn't be a peak. The chicks are dribbling out. Perhaps they'll keep on dribbling or, if the weather continues bad, they may all just go, and then there will be none left to weigh. Fledglings ready to leave are fully developed, plumage complete. Most Humble fledglings this season are on a staging post, a snow bank up from the two big colonies, giving them water to drink, somewhere warm to stand in the cold and cool if it's warm. They wait for their parents to return in the evening. Once parents stop feeding them, they lose 50 grams a day in body weight. At some point parents will make the decision to stop feeding the late chicks. Bill: It's hard wiring. They don't do it indefinitely, and chicks hatched too late in the season will perish.

From the staging post the fledglings go down on to the beach: they decide when. Then – they leave, and that's it. If you've missed it, you've missed it. You can't get them back.

Getting the weights is critical. Bill assesses possibilities against experience and decides we can slide in our light, nippy zodiacs over the top of flattish ice pieces and manoeuvre between the large lumps. So we pack our lunches and organise our gear. It's raining – not aggressively, but a continuous drizzle – and I add another layer, cursing the enforced awkwardness of bulk. I'm given a Rite in the Rain™ notebook, which means I can write in the rain.

We crawl and wriggle, reverse and shove, engines revving, propellers slicing like food mixers through the brash, needing twenty minutes to get through 300 metres. Approaching any island, I look longingly at beaches but we always land on rocks, leaving the sloping shingly shores for penguins. At Humble the landing rocks are high. We clamber up, pull off survival gear, fold it against the drizzle, weigh it down with stones and walk across the island to the main beach, an enticing wide curve between bookend boulders, low cliffs on the right, a steep penguin path on the left. A young elephant seal relaxes just above the tide line on red seaweed. Motley groups of penguins stand around: adults including moulters, fledglings, a few fluffballs. Some of the fledglings still have tufts of disengaging down. A bit remaining, a topknot of down, is OK – but nothing below their shoulders. Flipper fluff, front bits and behind, butt fur, neck band and pit hair – as defined by Donna – all mean their owners are still being fed by their parents and are not ready to depart.

Bill counts the fledglings and is happy: 110. That's a good score this year. Today could be the peak, but, data-cautious, he hedges. It might be tomorrow or Monday. I'm deployed on the left to stop the birds escaping up the path, Tim guards the right flank with two bamboo sticks. The rocks are slick with rain-wet guano. 'Stand still,' instructs Bill, 'then the birds don't panic.'

Don't worry. I'm standing still.

'If they are fuzzy, let them go. We only want the ones that are fully fledged. Any with a green spot, let them go. They've been done already. If the whole herd come at you, gently prod the leader at the neck with your bamboo and push them back.'

Bill decides to weigh thirty-five, a third of the total. Heidi and Brett wait in the middle of the beach, penguin nets poised. 'It seems like pandemonium', calls Bill. 'And it really is.' The first lunge yields one struggling fledgling for Brett, one to Heidi. Bill has a bird dangling in each hand. Penguins flee across the beach, bunching, then scattering. Bill is everywhere, long legs racing across the shingle, dodging the elephant seal, rounding up like a sheepdog. The penguins break for the sea, and Bill goes straight in after them, boots wet, trousers wet, heading them off. Captured fledglings are dumped into a holding box well above the tide line. Their flippers flap and clatter against the plywood sides.

The birds have clustered at my end of the beach. 'Stop,' calls Bill, 'we've enough.' And he helps stand guard.

The fledglings look up from inside the box, the distinctive white chins and staring ringless eyes of juvenile Adélies, thin heads and necks, greyish tinge to their new feathers. Their open beaks point up, like the nursery rhyme, four and twenty blackbirds baked in a pie. Heidi plunges her arm in, picks one out by the base of a flipper, grabs both feet and holds it in the swimming position while Brett measures culmen length and depth with callipers, then the length of a flipper with a ruler from the round knob at the base near the 'armpit' to the tip. Figures are called in millimetres for Bill to write in the notebook. The fledgling is marked two times with a blue cattle-marker and put into an adult weigh bag with a 5,000-gram scale. Heidi tries calming the bird's violent flapping, holding it through the mesh. As soon as the scale stops jerking, the weight is called to the nearest 50 grams.

'1,750.'

Bill: 'That chick is not going to survive.'

Released on the upside of the beach, the lightweight scuttles back towards its colony. The next fledgling goes through the same procedure. The weight is called. '2,350.' Bill: '1,000 grams below normal.' A penguin scrambles up on to the backs of its companions and escapes out of the box.

The next fledgling is streaked with fresh guano squirted in the crowded box. '2,800.' Bill: 'You can see what very thin chicks these are. The fledglings are seriously underweight this year.'

And I can see. Very quickly I can look at the thin, grubby bodies, pale brown-pink feet scrabbling upwards, beaks pecking, heads darting, make an estimate of the weight and not be far wrong.

The next fledgling is solid, squawking like an adult. '3,550.' Bill: 'Normal.' The defining moment.

Each fledgling on the beach has already done well. It has chipped itself out from a successfully incubated egg, been kept warm by two co-operating parents, been consistently supplied with food from the ocean and guarded from predators. It has moved away from its nest – extended its territory – crèched, chased parents for food and achieved the beach. Each has grown from a helpless chick, small enough to fit inside the curve of my thumb and first finger, to a fully grown fledgling the length of my forearm. Increased in weight up to forty times. Come through all the hazards, survived this season's harsh conditions, to reach this final point, standing here by the ocean, new feathers in place, with a blubber fat pad for insulation from the cold sea water; hungry, parents no longer the suppliers of meals. It has achieved a major goal.

In one sense all the eager fledglings struggling in the box are through into the next round. All are hopeful entrants at the start of a new life. But now, at the point of exit from the local island, from upright terrestrial life to marine life, what matters – the only thing that matters – is the weight that each fledgling has attained. The condition that it has achieved.

It's a simple as that. And as brutal. Failure, according to Bill, is already largely determined. Data from the 11,000 fledglings flipper banded on Humble have revealed the correlation between fledgling weight and survival chances, between caloric resources and blubber thickness at departure and the likelihood of return, to start breeding. Between potential success, or being fatally compromised. Every fledgling, Bill says, weighing 3,200 grams and over has the chance to survive life in the ocean, and return as a young adult to the islands. But fledglings that are too light lack the structural resources to survive the years ahead.

The box is empty. The remaining penguins are wary. The first grab scores four, then suddenly the birds are all in the other corner of the beach on the dirty snow, and the box fills up with flapping, fighting fledglings, thumping and jumping, some still cheeping, some half-way to adult voices.

Heidi: 'Who's next? You are – uh, oh.' Consternation in the box, all the fledglings staring up, fluff flying, guano spurting. A fledgling escapes into the sea and swims like a little duck, head up, so light its body is generally above water, diving then reappearing, flapping its wings, wagging its tail, dipping its head under, then running hurriedly out. Bill: 'This year is different. Usually fledglings go into the sea together. The leopard seals wait, scooping them up like popcorn. This year they are going in in dribs and drabs. And the leps don't seem to be here.'

At the third grab the last birds scatter, half to the left, half to the right. Weighing over, five fledglings only are left on the beach with a few adults. Tumbling down the rocks beside me comes a food chase: the adult running fast, well-fed chicks stumbling after. The smaller chick falls onto its head, rolls, picks itself up and they chase on over the shingle. Normally the beach would be full of penguins. Normally there would not be chicks here as young as this.

Skuas fly over. On the path up to the colony a chick lies, red sockets where its eyes were. 'Skua larder', says Bill. First its eyes would have been pecked out. Now it is dead, the skua will come back when it wants food.

There are no fledglings on Humble's second, smaller beach, but five on the back beach, all weighed but not measured because there's no box to pen them in. Back at the mooring rocks we find a flat surface with a back-rest for lunch. Wet, by definition. The drizzle has been continuous, with intermittent rain and sleet. I look out on the ice-covered sea, eating my sandwich, and briefly contemplate how other people are earning their daily bread, and think, I'm happy. Truly content.

Bill decides that he and I will go back to Station while the others continue with the work schedules on Cormorant and Christine, beyond the brash. Tim, selected as a leading US Boy Scout and visiting Palmer as part of his Antarctic experience, has been asked if he would like to hold a penguin and be photographed, and suddenly he has been initiated and will help. Bill attacks the ice in our zodiac like an ice-breaking speedboat: fast decisions, no hesitations. Sometimes we ride across the ice. Sometimes he backs off and rams. We pass a crabeater seal lying on a floe. A skua lands on a little piece of ice next door and stands there, on its two strong legs. Me: What's it doing? Bill: Perhaps checking if the seal's alive. Or checking for poo. Or just curious. They are very confident birds.

I'm meagre help harnessing the boat to the mooring wires in the heavy ice. But we are in time for Saturday afternoon scrubbing and sweeping, followed by a station meeting to hear and discuss the week's station business. Bob was up at 4.30 this morning to check an emergency in one of the systems. Palmer doesn't have a night watch rota – someone doing the rounds, checking visually. Instead, automatic dialling calls one person in rotation if there's a problem, and last night it was Bob's problem.

Donna's response to station concern about today's field trip is privately robust. We make a judgement. The work is manageable. If we can't get back, there are two alternative routes to station, both over land. And if we need to be rescued, the SAR (Search and Rescue) boat takes an hour to get ready. What is an hour? The wind has picked up a bit, compacting the brash, the 'white monster', as Jeff calls the raft of ice tracking about the Antarctic Peninsula, driven by winds and currents; the mother-in-law, because you can't escape it – and you never know when it will come. Brett, Heidi and Tim take a long time to work their way home. We watch as the three little red-coated figures inch slowly back across the jumbled white to tie-off. They are very tired.

At Amundsen–Scott South Pole Station the US Antarctic summer season is over. It finishes for the Americans at McMurdo Station, on Ross Island, next week. After his attempt to reach the south pole in 1909 Ernest Shackleton just managed to struggle back to Ross Island across the cold, unforgiving ice of the barrier, the Ross Ice Shelf, that mighty desert of glacier outflow, the world's largest ice shelf, reaching the coast on the last day of February. In 1912, having achieved the pole, Roald Amundsen believed he should be off the Barrier and home at his hut on the edge of the Ross Ice Shelf by the end of January. He made it with five days to spare. But on this day, 16 February, in 1912, Robert Scott and four companions were working down the Beardmore Glacier, from the high polar plateau where they had reached the pole a month after Amundsen. The whole lonely distance of the Barrier stretched ahead, cold unforgiving ice still to get across.

Even here at Palmer, in the Antarctic Banana Belt, 1,500 kilometres further north than Ross Island, summer is beginning to end.

On Sunday, Jeff is up at 5.45 a.m. checking our zodiacs. The bow lines tell him when winter begins arriving, he says. As they dip in and out of the water they grow a coating of ice, like tallow around the wick of a candle.

There's no wind, no rain, no snow, but Bill has declared Sunday a day off from fieldwork. Yesterday was tough, and everyone is stale. In ten days the *Gould* will arrive and remove a swath of summerers, including Brett and Heidi: the nominal end of the season. Some of the support staff can't wait to go. But science people are here for briefer summer slots and are feeling pressured, desperate to get their work done. There's a real edginess. 'Events' are being organised, an art show, a variety show, the Palmer special – a one-night appearance of the Neanderthal Café, dinner for cavemen and cavewomen served by cave drongoes. All high-action Station activities, but the same few seem to do a lot of the organising. This morning I'm running a writers' workshop, but I really

need to write up yesterday's field trip. Bill has to get through accumulated administration, but he's torn. Given this unusual season, he wants
to be out on the islands, looking at what's going on.

Support staff enjoy the day off. People sit in the galley chatting
and producing yet more woolly hats. This season knitting, creative and
relaxing, is in favour. It's social time, but not focused on the bar, which
causes some resentment. Being Sunday, Wendy makes bagels as a treat.
Otherwise, as on all weekends, we get our own meals out of plentiful
leftovers. It's a chance to revisit earlier delights – Friday night's ragout,
Saturday lunch's salmon baked in a cheese and spinach sauce – each
tempting meal just a choice, and a microwave, away. Everything not
eaten will be thrown out on Sunday night. Wendy and Jennifer create
'home cooking', California-style, for their summer clientele: delicious
and inventive given the quantity and quality of fresh fruit and vege-
tables in daily decline until the next ship brings new supplies. Brit-
tney, the Raytheon 'Waste Technician', is making a recipe book of our
favourites.

Bob is tidying up. He finds a box of old T-shirts and hands them
out. Someone comes in and says, tensely, is there one for each of us?
The culture of sharing comes with making sure you don't miss out on
whatever's going. This is the third time I've come to Antarctica funded
as a writer. I recognise some of the pitfalls and generally keep my head
down.

Monday, rain, rain, and Bill returns to Humble for the sixth round of
weighing, counting seventy-five fledglings on the main beach. Numbers
are plotted in a curve, the peak date being the maximum. In a typical
year fledging lasts from 2 to 23 February, with a peak between the 10th
and the 14th. Now, last Saturday, the 17th, looks like being peak fledging. Numbers appear to be curving back down again.

The last of this season's tour ships is in, a Russian ice-strengthened
ex-polar research ship, with mainly Australian passengers, so I join
them for brownies in the galley and find friends of friends, and friends
of relations. People on Station laugh because, of course, an Australian
would know someone among any group of Australians. Later, by the

landing rocks, the expedition leader with the trademark battered rabbit fur hat looks across to the mainland, where the mountains have suddenly revealed themselves, checks the wind and declares excitedly that the evening will be clear. Expeditioners hurry to board, eager for their next adventure.

Everywhere on Station there's the release – the relief and happiness – of a beautiful evening. People move fast, capturing long-awaited photographs, or just stand quietly, looking. The ice cliffs accept attention, not glaring, repelling the gaze, as in full sun. In the transparent light the islands gain a solidity of outline, a sense of certainty about their exact position. In mist, rain, snow, the islands become outlines, loom forward, recede, are blanked out. Now even the distant islands are clearly claiming their place on the horizon, brown rock forms against the paler silver-gold sheen of the water. Flagstones and pebbles of ice float in the sea interspersed with small pieces of sculpture, and those whimsical mini icebergs, fretted and curved, water-worn remnants of something greater, with additional fragments frozen in where they happened to attach, arbitrary pinnacles, angled superstructures, stuck-on cubes. The sea is peaceful. There's a sense that the ice is already exercising a calming hold across the swell, like an invisible silken surface.

We watch the sun set. It no longer hovers and lingers, but sinks quickly, conventionally, down into the ocean's horizon. Tonight there's a crescent of moon visible for the first time and, at last, stars. The skies have been too heavily clouded to see them till now.

On the way to bed at 11 p.m. I find the Southern Cross, that potent symbol of home for Australians. The two pointer stars are pointing due south, unimpeded. Keep on going, a little to the left of the Dive Locker, and there's the south pole.

25

Not a standard day

Thursday 21 to Tuesday 26 February

This morning the Carp Shop is filled with fish, the high airy space become a visual aquarium. Images of Antarctic ice-fish lie on every surface, their bulky ice-fish heads with big round eyes and wide receiving mouths, their curiously tapering naked bodies, transformed by coloured inks into prints on art paper, shirts and redundant white cloth from the medical stores.

Last night we took their dead, intact bodies, held since winter in a lab freezer, and laid them into excavated blocks of polystyrene. The fish looked just as they did when they came out of the ocean: the detritus of their living still attached, their bodies as platforms, globular yellow copepods sticking to their fins and bellies, translucent white copepod egg lozenges lying beside. The next cycle of minuscule crustacean life in waiting.

We came to the Carp Shop direct from a lecture on marine leeches. Gene, the only and therefore the world taxonomic expert on marine leeches, examined our soon-to-be art and, face glowing with pleasure, found a specimen of an Antarctic marine leech. Then we painted the exposed flank of each cradled fish with colours, subdued and subtle, or gilded like tropical rainbows. We could choose. Each of us with the sudden privilege to anoint the dead. Placing cloth or paper over the form of the fish, we felt through our ignorance to the hidden shape beneath: the structure of the jaw, the rise of fins, pre-pinned open, the splay of the tail. Pressing gently back and forth, transferring the image from body to shroud. Each time the anticipation of discovery. Each time, lifting the covering, the wonder of seeing which part of this once living animal had been revealed. Each fish giving up a particular aspect, an unsuspected emphasis, of itself to a kind of posterity. Sometimes the round eye stares. Sometimes the predatory jaw dominates. Some fish seem to swim, some hunt. But most of all they look, to me, like fossil images, the shapes of ancient fish trapped inside rock.

When we finished, the fish were washed free of all polluting paint and put into the sea. Nine-tenths of all the fish species swimming in Antarctica live nowhere else in the world. But the most startling of the meagre list of Antarctic fish species are ice-fish. Their blood is no colour. They have no haemoglobin in their red cells. They manage the sub-zero temperatures by carrying oxygen dissolved in their plasma and anti-freeze in all their body fluids. Ice-fish can live a long time, an adaptation of many polar animals to cold. Life pacing to a slower beat: years as a juvenile, maturity and breeding beginning later. Population numbers easily and irretrievably disjointed by premature death. In just four years of fishing around South Georgia the biomass of fish almost halved. Our fish were channichthyids, caught as part of a winter trawl by scientists who needed samples of their interesting blood: an inter-vention leaving their forms intact. Extras in someone else's drama.

I scavenge my fish images from the Carp Shop quickly. Donna wants a 10 a.m. departure on what she calls the usual protocol run. It's a chance to see what needs doing, what's there, at the start of February's final week.

Humble first, for fledgling weights. There's a small cluster of birds at the far end of the main beach with moulters, a few late puffballs standing tight, and twenty-two that Donna estimates can be counted as fledglings, on their way out the door. Heidi and Brett estimate higher, resulting in a discussion of that defining characteristic, amount and position of fluff. Several fledglings are 'stale', blue stripes indicating they've been weighed before, and one – still with cap and flipper fluff – shouldn't have been. Most are 'fresh', unweighed, and Donna decides to catch eight, avoiding upsetting the moulters if possible. 'Timing is the essence of catching fledglings. I keep emphasising, go slowly. Slowly. Then nab them. Don't rush it. If you do, one chick runs like the head of a spear and the others follow, then starburst out, breaking in different directions. You will lose them. The chicks get riled up. Sound travels through stones, and the beach is shaking with sound.' The wooden holding box is carried closer to the penguins to avoid the resident elephant seal. In the general grab they do include four birds still in tufts and topknots, but the rest are rushing back across the beach or up the rocks behind. There's only this one chance to nab. Donna has two birds in one hand, a third in her other, a fourth under her arm, agitated birds pooping on her clothes.

Fine down and little bits of guano hover in the air. When there's big numbers, we gag and wheeze in the cloud hanging over the birds, says Donna, measuring and then writing in her own notebook. The sun comes out, and the rocks, camouflaged with last night's snow, start glistening. A bird runs fast, newly adorned with today's colour, green, stumbling over the pebbles. Where does it go? Donna: 'It looks for the last place where it's seen its buddies.' The batch are lightweight, though better than other days.

The second beach has no fledglings, nor the third: only twelve moulting Adélies and one moulting gentoo standing on a snow bank. Peak fledge date was last Saturday, as Bill calculated. Fledglings leaving Humble main beach form a bell-shaped curve; but this year the line has staggered, peak followed by a little shoulder. The pitifully low number of chicks has affected the pattern.

A pair of unbanded brown skuas, arrived on Humble last season, have bred for the first time. The nest-sitting male is removed for left-leg banding, but the angry female – at least, the strong guess is that it's the female – avoids catching. Fourteen days ago the chick was an egg. Now it is weighed in the smallest weigh bag, pausing briefly in my curved palm, a tiny grey downy 66 grams, before being tucked back under the parental wing. The two chicks of the skuas that have guarded the entrance to Humble all summer are running over the rocks like long-legged hens, or mini short-legged emus. One parent sits on a rock, the other stands near. What are they doing? Hiding from their chicks, Donna says enigmatically.

Brown skua territory boundaries on the inner islands are currently adjusting for the second time since the 1970s, when avian cholera swept through the Palmer area, hitting the browns and eliminating some historic pairs. Torgersen became vacant, but the brown pairs owning Humble and Litchfield managed to survive and control that resource as well. Now two new pairs from the outside, so with no history, have arrived on Torgersen and succeeded in producing chicks. Except for Litchfield, the browns on all the penguin islands have produced chicks this season: Bill says that, on average, brown skuas do better than south polar skuas.

We cross to Cormorant through six-foot swells and a choppy sea, Donna wedging herself across the zodiac to minimise painful bumping on her backside. The snow bridge has collapsed and more of the long-term snow has melted out, revealing the island's attractive small-scale angles and local views, little headlands and rocky islets. Cormorant has a excitingly close view of the glacier face and Mount William. Sometimes I dream briefly about this island as holiday paradise – and banish the thought. Animals and place are in fragile balance.

Fledgling numbers are so reduced this year that Bill has adjusted the protocol to include weighing chicks on Cormorant and Christine. But they are being weighed while still in the colonies, in advance of their departure, because so few chicks are actually making it to the beach. We find fourteen almost-fledglings on a small saddle above an enticing cove, some already streaked with blue paint. Donna grabs three in a net and one under her arm. Brett and Heidi have another three. An uncaught fledgling jumps up on a flat rock, watching us. Its left wing has not formed fully, nor its left foot, but it can run and keep up with the others. The Birders call it Lefty. Ambushed by anthropomorphism, I think it looks particularly alert. 'Poor little critter,' says Brett. 'It can't survive in water. It will only be able to swim around in circles.' Yet here it is, making it through to the next round.

Three of the newly weighed birds rush fast down the slope, sledging on their bellies. A fourth runs across the rocks. It stumbles, then staggers. Donna picks it up. Its head droops. She lays it gently on the ground, facing in towards a low shaft of dark rock. In three or four seconds it is dead.

The weighing over, Brett picks up the dead fledgling. Its stomach is almost empty. On weighing it was 2,100 grams. Two days ago the average weight of the chicks weighed here was 2,200 grams. The average weight for the seven fledglings weighed just now is 2,500 grams, so the dead bird is 400 grams under that average. Brett puts it in a blue bag over his shoulder, to take back to Station and dissect on a parasite hunt. Then it will be given to a small college and become a skin: a rare example of a dead Adélie fledgling undamaged – not dismembered by skua or giant petrel, not bitten and flayed by a leopard seal, not failing somewhere at sea of starvation – but dying on the rocks suddenly, in front of us. Malnourished. Donna says that she has never seen this happen.

A chick remains on the saddle, and Heidi chases it to the beach to join its mates. Staying here alone is too dangerous.

The brown skua chicks on their nesting platform are checked, their flight feathers measured. Donna demonstrates how the top point can be found, to start the measuring, without damaging the feather. Figures are noted down, compared with the previous readings, checked for general validity. An anomaly is pointed out as an interesting possible outcome, and the feathers are remeasured, resulting in new data. The knowledge of experience. The experience of handing on knowledge.

The cream-white guano on the way to and from censusing the blue-eyed shags and their by now elongated chicks is so slippery after last night's snow that we balance on embedded rocks. Suddenly Heidi hears a faint cheeping. Two lichen-encrusted grey slabs lean together on a base rock forming a small triangular passage, and, peering inside, we see one Wilson's storm petrel chick and an addled egg. Rats have jumped ship and established themselves on some sub-Antarctic islands, and cats infest others. But here in Antarctica birds can nest on the ground or among the rocks, free from marauding mammals.

Cormorant Island has an abundance of living: blue-eyed shags clinging to their creamy cliffs, Adélies in their scattered colonies, brown skuas guarding their territory, kelp gulls, giant petrels, elephant seals, fur seals, today a Weddell lying prettily on a snow bank, blotchy silver-grey and big lustrous eyes. Visible life, plus the microscopic inhabitants of nest stones and moss, and the usually invisible fellow travellers. At Colony 5, where every chick is weighed, Brett notices a long, yellow-brown tapeworm emerging from a chick's only rear exit. Delighted, he tries to remove it, for study. Except it won't oblige. 'You'll have to deal with that yourself', says Brett, putting the chick back on the ground. Then, generally, 'They are infested'. Worms infest fish, infesting birds and seals.

We cross to Christine, with Heidi and Brett jumping off the zodiac as it rises on a swell to scramble up a rock face and check Colony 1.0; but it is empty. Eighteen days ago, when I was here, Colony 8 had one larger and one smaller group of penguins, separated by an unoccupied no man's land. Their silhouettes, sparse along their ridge top, stay in my memory. Now the spaces are empty, and will remain empty, until next October. Which, if any, of the chicks on the beach are from here? It is not possible to know. I can have the census data. That is all. There can be no record of how many, if any, of the chicks raised on Colony 8 achieved a viable weight. Or which predator killed which chick – whether with a defending parent, or in a food chase or alone, vulnerable, on a path or the beach.

Donna moves fast, despite the bruising. All chicks at Colony 6 are weighed: one grubby little bird is only 1,250 grams. Water lies between the rocks, shallow pools of pink-brown liquid. Slick rocks, slippery guano. Was it this wet when the snow melted in early summer? Heidi: No. Never this wet. At the next colony Donna gets four chicks in a net, then with one between her legs, one over her back, two in one hand, begins weighing, and writing. But her five-chick hold is the ultimate: one between the legs, one tucked under the upper arm, two hanging by a single flipper from one hand, one from the other. Everyone is preoccupied, not stopping. Watching our steps on the slippery rocks is essential. And it's cold. There's no pause for food: fossicking out a quick snack is all there's time for. Birders have their own non-issue clothing, with big pockets in sensible places. In the

field everything needs to be to hand: stacked in the pockets, predictable, in order, like the most efficient filing system. This year my issued
clothing has inadequate pockets. I'd put the right pockets ahead of
almost anything.

Christine's brown skua pair are long-time residents. The route up
to their abode lies through ante-rooms scattered with the remains of
meals, the mossy gaps between the rocks littered with fragments of
Adélie chicks and now fledglings. I pick up a complete little foot, three
toes with dark toenails, soft pinky-brown skin, webbing between. A
fledgling head lies on the ground, a ruff of white feathers around the
neck like the ruff of a discreet seventeenth-century Dutch merchant.
The backbone trails off, empty of most flesh. The progression of growth
denied, to create new growth. All a bird's fat and protein are interior. Skin, feathers and bone are indigestible, zero calories. But the carcasses represent the expected predation. Nothing different this year. As
Heidi reminds me, the maths is obvious: 100 penguins, 20 eaten and
80 survive; but 50 penguins, 20 eaten, means only 30 survivors. The
Birders' interest is in the visible part of the population. The number of
survivors is what matters.

Brett and Heidi catch a fat, healthy brown skua chick each, and
sit on the rocks measuring culmens and flight feathers, Donna sug-
gesting, demonstrating. Both chicks are now big enough to be fitted
with lightweight stainless steel bands. The leg must be stretched out
in a natural angle. The band must slide up and down in a circle, not
an ellipse, and be placed the right way up, so the numbers can be read
from the top down. Donna takes a chick, demonstrates the hold, the
way the leg is presented. Brett and Heidi must remember, remember.
I'm remembering what I am seeing – they, what they must do. But
there's still so much work remaining that Heidi has agreed not to leave
on the *Gould*.

On the track back towards the beach, a freshly killed fledgling lies
on its side. Nothing will take it. It won't rot. The skuas will carry it
off when it suits them, tear away the flesh when they need it. Some
penguin remains on the islands belong to previous seasons. At a large

Adélie rookery in eastern Antarctica I saw the ground littered with the bodies of chicks in their grey fluff, flattened and desiccated but intact, undamaged. How many years had these little glove-puppet bodies lain there? Nest-building, egg-sitting, went on all around, the annual business of creating a new generation.

Heading home, Donna notices a wedge of fine spray towards DeLaca Island. High swells could be hitting the upturned hull of the wrecked *Bahia Paraiso*. We stare, and then we see another wedge of spray, then another, not in the same place – and know the quiet exhilaration of sharing the bay with whales. We drive across the bouncing sea and stand by, riding the swells, watching four humpbacks feeding. Sometimes we are close enough to hear the exhalation of humpback breath, smell the sudden fishiness, watch one mighty back rise in a slow long curve, the small central fin riding like a rudder in reverse, see a second back rise and disappear slightly behind. They breach, blow and dive deep, showing their white and black flukes, while we balance in a small shallow dish on the skin of the ocean. Last time I was at Palmer, people from Station went out one calm evening in the big zodiacs to watch two humpbacks cruising slow, relaxed, in Arthur Harbour. Suddenly they turned and swam straight towards us. I knelt in the zodiac, leaning over the side as the solid grey back, the enormous flippers, the white mottling on the body, passed on and on directly beneath. Twelve metres of humpback in the calm water. Now the surface is wreathed in spreads of white foam from churned-up phytoplankton, and little balls float where two foam spreads have rubbed together in the swells.

Tracking the whales, we have moved further north, so Donna changes plans and decides to tour the northern islands, showing Heidi and Brett where to look for kelp gull nests from the sea. It's new territory for me – handsome crumbling ice cliffs, bare rough tough brown rocks of islands, a long barren promontory. We swap treats. I pull my London supplies out of a pocket, Kendal mint cake, and a slim packet of the ultimate chocolate, By Appointment to Her Majesty the Queen Mother, made in north London, 99 per cent cocoa. I break the chocolate off in small tablets and explain – let it dissolve in the mouth. A little goes a long way. Donna gives me jerky, local Montanan, made of elk. A little goes a long way.

Last night's storm, combined with a very high tide, has pushed pieces of ice high on the rocks. They pause, poised, till tonight's tide, which might catch them and bring them down. Some of the coves are crammed with ice, holding bays for transient cargoes. One piece, water-smoothed, is clear with subtle tones of blue, like finest-quality Lalique glass. It balances, tide-placed, high on a flat rock. An art installation. We capture two lumps of clear glacier ice bobbing past and haul them into the zodiac. Bar ice for drinks. They lie by my feet near the Adélie fledgling. Strange to be so close. Strange for the fledgling to be so totally still. I know it's dead, but I've been looking at these birds always in movement, or as dismembered skeletons, bits and remnants – but never complete yet completely still. The myth of funny little penguins, when in truth their lives are tough, driven, unpredictable.

Donna: 'This has not been a standard day.'

We are back in time for Station meeting, a day early, and the Station photograph, everyone in orange survival suits squeezing into the zodiacs. Heidi and I clean the labs, the House Mouse job that's been reserved for the Birders; there's gash duty for me, then a trip out in the big boats because the evening is beautiful, and I start writing up what, even for the Birders, has been a big field day. Donna, Heidi, Brett and Bill go to Humble for the season's last diet samples and find only three penguins on Humble to lavage. But I'm not being as sharp as I need, so I go to the lab. The dead fledgling is lying in a tray, the body still pliable. The feathers on its face, its yearling neck, are white and very soft, the white chest feathers densely overlapping. The black feathers on the back are short and thick, like thatch. One tiny bit of grey fluff on its back clings, wispily, to the end of a new feather. A feather seems to share the same pit as the down, pushing up and out, carrying the down along as it grows. The leading edge of the flipper has a fine, parchment-like scaliness. The bill is small and dark. When Brett dissects the fledgling, it turns out to have very little fat, with no krill or fish parts in its gut. It is deemed to have died of dehydration, and emaciation.

The air doesn't smell in Antarctica – but now it is warm enough, and it smells wet, and fresh. Stormy winds continue day after day, with snow, rowdy seas and rain, limiting or preventing field work. The *Gould*, working south in Marguerite Bay, has to run for shelter, the weather is so bad. Whale researchers on board can't even do visual surveys for whales. Almost forty years after hunting blue whales, *Balaenoptera musculas*, ceased in the Antarctic, there is currently real uncertainty about their numbers, and the subspecies to which they belong. No sightings of blue whales have been made off the Antarctic Peninsula for several decades. The researchers are attempting to find answers by using acoustic recorders, a navy device for submarine detection. A year ago eight acoustic recorders were left as moorings on the deep sea floor along the western coast, continuously recording whale sounds. This is the pick-up cruise, and the data being retrieved, they say, is rich pay-off. Blue whale calls have been detected all through the year. And it appears that the same subspecies of blue whale exists all around the continent.

By Tuesday 26 February there are no fledglings on the Humble, Cormorant or Christine beaches. They've all gone. In a normal year 100 to 150 fledglings will stand at the water's edge, calling and calling. Donna: 'You can feel the momentum before they hit the water. The potential energy. The build-up. The drum roll. When a big group goes in, nothing is going to stop them. It goes kinetic as they enter. In the end it's an amazing display of instinct. It's "I need food." But this time it just happens. It's quiet. One or two chicks may peer in together. They are more cautious when alone, hesitant. One chick might approach the water and jump. One might follow an adult.' Bill: 'The giant petrels aren't waiting in the water to pick them off. The leopard seals appear not to be around. The chicks are leaving in dribs and drabs.'

Bill's 'light bulb moment' about climate change as a determinant in demography occurred on Humble in 1995 as he weighed and banded Adélie fledglings and realised that the chicks were underweight compared with past years. The chicks had been hatched late in the season in a year of heavy snow. As has happened this year. Mathematics defines

15. *This satellite picture shows Larsen B Ice Shelf disintegrating, ice milling in flotillas of clotted bergs and brilliant blue brash. Image acquired by the MODIS sensor flying on the Terra satellite on 17 March 2002 – a clear sunny day across the Antarctic Peninsula. The peninsula's west coast opposite Anvers Island is visible top left. See map on page xi for the current configuration of this part of the east coast after the departure of the ice shelf.*

16. *Adélies and chicks on Cormorant Island, with close view of the ice cliffs and Mount William.*

17. *Leopard seals eat krill, fish, other seals and, especially during the breeding season, penguins.*

18. *Ferocious summer Adélie chicks near fledging. Chick on the right has been weighed and marked, chick at the back is stained with guano.*

19. *This giant petrel, hatched in January 1994 on Norsel Point and began breeding on Humble Island aged six with a slightly older female from Stepping Stones, is seen here tucking his nine-day-old chick under his body. During the ferocious summer the pair lost their egg when their nest flooded. This little chick, a male, hatched in January 2003 and fledged four months later.*

20 and 21. *Norsel Point on 15 January 2002, still joined to Anvers Island by the
ice of the Marr Ice Piedmont. Base N and Old Palmer were about half-
way along the south-west coast. The final ice collapsed two years later on
10 January 2004. Bottom picture shows Norsel Point as an island in
January 2007.*

22. The author writing on Limitrophe Island, orange-colour lichen on the rocks.

23. *The beauty of the inner passages on the Antarctic Peninsula's west coast: the entrance to the Lemaire Channel, 'Kodak Alley', between Palmer and old Faraday Station, now Vernadsky.*

the reality of each season. But a range of circumstances determines the proportion of fledglings that achieve a weight above the hurdle. Bill summarises: this year late heavy snow delayed laying. Continued heavy snow affected chicks' survival chances. Late hatching meant parents were foraging later in the season. Again the model predicted the outcome, given the weather conditions. The impact of climate warming: the landscape effect – it's all grinding together in a horrible reality. The final fledging weights are not yet in, and Bill will wait for his numbers. But this season's fledglings, the window of their opportunity reduced, are underweight. The hypotheses are being proved. The natural experiment is playing out. It is delivering the results.

Except there's a new wild card: the ferocious weather continuing on from November and December into January, and now the whole of February. Nothing like it has ever been experienced at Palmer. It's totally atypical. Repeated storms belt in, with barely a pause, high winds, some snow. But the temperature is warm enough for the moisture-laden air to unburden itself in sleeting, driving rain. All the things I'd ever read about the Antarctic continent – the highest, windiest, driest place on Earth, the coldest, with most of the planet's fresh water held as snow, ice, glaciers, ice sheets – had never mentioned rain. Rewrite the books. This summer at Palmer it really rains. As if I needed further proof today, wind gusting to 50 knots, rain squalls sweeping past, I see the never-to-be expected: an Antarctic rainbow. It spans Arthur Harbour, one end on Bonaparte Point, the other anchored by the ice cliffs in the hump of atrophied glacier slowly shrinking above the soon to be released new island. A potent symbol.

Bill's models predicted what would happen if reduced numbers brought Palmer's Adélie colonies below the safety line. This season the impact of early storms on egg, and chick, survival, the departure of failed parents, all knocked the size of many colonies perilously low. Predators – brown skuas, and now giant petrels – picked chicks off one by one. Weaker chicks, late chicks. Bill: 'The colonies were reeling. The structures collapsing.'

And there's that other, highly significant, factor. Around half the adults didn't even arrive. This season Adélie numbers started low, and fewer losses were needed to trigger disaster.

Looking for answers leads straight into the ocean: what happens to each penguin when it leaves its breeding colony? What happens at sea, and especially, during winter? The vast world of Adélie unknowables.

Palmer's Adélies are a precise, manageable story of climate change. This season is providing the proof: hypotheses answered, data in place, bits of the jigsaw nicely slotted in. Now a loose end inserts itself: adult Adélies fail to appear on the Palmer beaches at the start of the season.

To me the missing Adélies symbolise a crucial aspect of trying to understand climate change. Follow the loose end and it will hook in more factors, insinuate into more networks, lead, at some point, to new understandings. Bill has already begun fieldwork, data collection, formulating and testing hypotheses. Ecology is a continuing process, season following season. So many variables are at play. Climate change means studying Earth's insufficiently understood systems, accepting their interconnectedness, their complexity.

Palmer's penguins are a micro-story: clearly, and painfully, warming's losers. But the matrix of ocean, atmosphere, ice and land they inhabit – that leads into the macro-story, all the global concerns of our warming planet.

The build-up to the departure of the *Gould* on Wednesday peaks with an orchestrated 'pie event' late Tuesday afternoon in the Dive Locker. In the middle of an orderly queue planting kitchen-made pie concoctions over a supine Dave Bresnahan, recorded on video, a Chilean navy ship arrives unexpectedly in the bay. Ceremonial horse-play is replaced by the speediest of clean-ups, and rapid hospitality. The grey ship, on patrol duties, moors overnight in Arthur Harbour, and officers are invited to dinner, Station tours organised. A naval vessel from each of the three historic claimants to the peninsula – the UK, Argentina and now Chile – has visited Palmer this summer. Dave Bresnahan: Maintaining a presence with three permanent stations, at the geographic south pole, McMurdo and the peninsula, and an influential science programme, means the US can play a major role in influencing Antarctic matters. Me: What are the political interests of the US here? Dave: 'Look on the damn blue cup in the station store.' So I buy one and read, among other pronouncements: 'The United States Antarctic Program shall be maintained at a level providing an active and influential presence in Antarctica designed to support the range of US antarctic interests.' The cup is made in China.

Science was inserted into Antarctica as an enabler, the means of resolving a mid-twentieth-century political impasse at a time of high military tension. In the first decades after the signing of the Antarctic Treaty scientific work in Antarctica tended to be routine, parochial, low-cost, scattered, with little effort to set the science in a wider context. But Antarctic science has developed its own powerful momentum, agendas and interest groups, independent of its originating role. During the 1990s Antarctic science realigned, to become a significant part of the world stage.

Scientific activities can be, and are, a convenient cover for politics. But Antarctica has so far been kept free of territorial claims going final. Science's role in this extraordinary achievement is vital, and barely acknowledged.

26

South polar skuas and kelp gulls

Thursday 28 February to Tuesday 3 March

A posse of fur seals commandeers the blue safety barrels, giant petrels have pegged out the small entrance promontory, but Shortcut Island's undisputed owners are south polar skuas. Bill holds David Parmelee's big old skua net, long handle, black square metal frame, wide mesh hanging like a monk's cowl, keeping it high against attacking south polar skuas. The rest of us hold up sticks of bamboo. This morning Heidi and Raytheon helper Rick are working the east. Bill, Donna and I turn right up a short pebble-strewn chasm to the west. Shortcut's rough rocks, its tumbled landscape, sit inside the Birders' minds divided into ridges A to H with precise boundaries, each subdivided into nesting territories with its own south polar skua population. Some have been at Palmer more years than Bill: a number of known-age birds nest here, KABs, banded around the right leg as chicks by Parmelee, giving a run of nearly thirty years' data.

Working a ridge we hear faint high cheeps, and a small box of rocks is discovered with a tiny chick still damp from the egg. For the log, for ageing purposes, this is a day zero chick. Despite chances of survival being pretty dismal, this day zero chick will have to be worked with. Bill: 'If we ignore it we are missing information. We will measure weight, length of the bill, and if it survives, length of primary feathers which start at day seven. The relationships between all three give us indications about food supplies. Bill length is a very sensitive indicator of feeding. But evolution has prioritised some things in animals to grow independent of food supplies, the things most critical to survival. Kelp gulls and skuas can't do without primary feathers, which grow at a very steady rate whether the animal is well fed or not.' The day zero chick risks chilling so we move away fast. A downy first chick is already in the nest. Both chicks hatched a month later than the usual date for Shortcut's south polars. But this summer south polars bred either late, or not at all.

On the next ridge a single chick with vigorously defending parents is measured and weighed. Five days ago the chick shared the nest in the sparse grass with a younger sibling. Now the sibling is missing. Five days ago this chick weighed 145 grams. Now it weighs 240, but it has relieved itself of a spurt of white guano, and regurgitated an impressive three lozenges like mini-breakfast sausages of dark pink penguin mince. I hold it cupped in my hands, webbed feet tucked up safely, little wings poking like small flippers out of softest silvery-grey down, while Bill scoops a parent into the net without too much difficulty. Donna holds it firm against her body and Bill aims a precisely timed finger pinch onto the strong weapon of a bill, measures depth and length, then the length of the tarsus, both indicators of sex in adults. Male south polars are a little smaller than females. Is it a male? Perhaps. Released from the finger pinch, the south polar darts its head back and forth, pecking angrily at Donna's blue jacket, squawking and protesting. It stares up at her capturing face, her eyes carefully protected behind sunglasses, then regurgitates into her all-purpose kangaroo pocket. The stainless steel band, numbered in sequence, is placed around the left leg. Bill brings the strong steel together with pliers and checks the band for loose movement. The number is noted and the bird released. Four to five minutes of intense, skilled action from disentangling out of the net to release back on the ground.

The second parent, curiously blond, wary, sweeps by out of Bill's reach, calling its evocative warning call. Donna crouches in the nest site making convincing skua noises, throwing pieces of moss up in the air in imitation of angry skua behaviour. 'Hold the chick up' and, kneeling low, I hold it high. The chick calls vigorously, but still the parent skirts Bill. A south polar club, unattached birds from which mates can be selected, stand in a line watching intently on the slope above.

Three have right-leg banding. Out come the binoculars, and numbers engraved into bands curved around the twig-thin legs are read, called and noted. One from the 1990–91 season (Bill: 'I banded that one') the others from 1994–95. Each number is a record of survival, the start of a life history; if a KAB breeds on Shortcut, it can be put exactly where it belongs in the data.

Suddenly the blond bird is in the net, being disentangled, held by Bill while Donna does the measuring, turn and turn about in this work except they swap holds while Bill clamps the steel band together. And after all, it is the male.

Donna is on her knees examining the male's regurgitation, which contains fish, so she picks out the jaw with a teaspoon, in the hope of getting the identifying otoliths, putting it in a named and dated ziploc bag. She scrapes the regurge out of her pocket, adds white guano scraped off the rocks, and picked out of the nest, trying to minimise the amount of moss. Poop and regurge samples are collected throughout the entire breeding season, frozen, then analysed next spring to see how diet affects south polar reproductive success.

The chick sits quietly on the ground as we back away. The penguin-meat sausages wait to be re-fed – visible food chain. An Adélie chick, sustained by two parents through days of foraging in the ocean, killed – who knows how – and scavenged by a south polar skua.

We range through each breeding territory and across the invisible boundary to the next, checking nest sites, peering into crevices and beneath ledges, balancing like careful giants on mini-ridges, by mini-canyons, hunting around cascades of moss and still ponds of fresh water held in sharp-angled rock troughs, always watching for the scuttle of a grey chick, the glimpse of a small, watchful head tucked under an over-hang. The landscape is complex, full of hiding places. We observe adult behaviour, and the state of nests, round shallow indentations in grass or moss. Sometimes the grass is thicker and brighter, or there are signs of activity – streaks of white guano and small airy balls of regurgitated penguin feathers, the white feathers of the breast curled into a bolus. Chicks that were recorded as present five days ago and cannot be now found are recorded as missing. Chicks that were recorded as missing five days ago and are still missing are written down as possibly dead. Again and again the chicks have gone. We are recording failure: failure to breed, failure to raise chicks. Shortcut this season is another disaster.

At a nest site in the middle of a beautiful moss-thick micro-valley, both parents defending with insistent vigour, we search and search but find nothing. Five days ago two small chicks were here. Up the slope a giant petrel lurches over the rocks, wings outspread to balance. Folding legs under its dark-plumaged body it sits, looking at us, then walks on diagonally down the slope, treading the territories of three or four nesting south polar skuas. I think it looks bold, and slightly shifty. Donna politely chides incipient anthropomorphism. It's young, she thinks. Probably uncertain, perhaps lost. If there's no wind, like today, a giant petrel can't lift off. It might have been eating a south polar skua chick five minutes ago. Or the last storm could have pushed the two missing chicks over the edge of survival. In a 'normal' season two south polars can see off a giant petrel in their territory, even bowl it over. They could, if they really wanted to, drive a human away.

Bill stoops and pulls the body of a last-season fledgling out of a narrow pool of dark water. Adults will take fledglings as they fly, he says.

Back on Station I'm shown the Birder Box. Bigger than a big upright freezer, made by the Commander, a legendary carpenter at a time when these things happened, it has wooden handles for lifting, and stencilled notices: 'S-013', 'KEEP DRY', 'DO NOT FREEZE'. It's a paradigm of the seabird ecology work. Every season, filled with necessities, the Birder Box was retro-ed to Punta, to be stored and brought back for the new season. Inside is a history of practice. A gun rack. A telescope monopod, a wooden feather box, with feathers still inside. Treasures and memories: an X-ray of a giant petrel chick to see if it swallowed a hook from a long line; a shark's egg found on the Joubin Islands; a couple of toys. Everything labelled. The interior, divided into shelves and bin spaces covered by plastic doors on hinges, is neat and efficient. There are spare gloves and mitts, socks, gaiters, towels and dry bags, day packs, hand- and toe-warmers, an extra wind-top, grubby and worn. A box with medicines, band-aids and skin care. Dissecting kits. Plastic weigh bottles, scales, goggles. Samples of pink guano in dated ziploc bags. All the lavage stuff. Taped inside is the list from the Punta Arenas warehouse, the retro form that has to be filled out, everything listed from 'Notebooks 16' to boot treatment to 'Telemetry Logger, 3, $7,500'. And at the back of a shelf, a tin of strong, weatherproof adhesive with, written on the underneath: 1. drink 2. uncap 3. huff 4. yippee. Heidi laughs: the world of Jeffco the Dancing Outlaw. Inner-circle stuff.

This autumn's lab dismantling means the death of the Birder Box. I speak out on behalf of Antarctic history and the history of science. This is a wonderfully coherent, concentrated, intact artefact. It should be saved for a museum, if possible. But that's logistics, storage, costs, decisions, time. Much better, apparently, to be rid of it.

The sun sets, but the sky is still a clear warm yellow-brown above the horizon, with an imperceptible merging into a brighter pale green-blue light which becomes the dark of the night sky. Looking east at the peninsula I see a bright light on a mountain's edge, like a reflect-ing mirror. A light is impossible here. Then I realise that it is the top edge of the moon, and watch it rise gently and evenly above the moun-tain, bright golden, until its reflection shines straight along Hero Inlet. Small pieces of ice ride in the water, white where the water is dark, black where the water is light – the pale green-silver water-light from the sky, still coloured.

A bergy bit floats in the inlet, the size and shape of a zodiac, with the silhouette of three people hunching slightly forward. But there is nobody. Except us.

Next day, 1 March, the sea is so calm it nibbles gently at rocks normally swept by swell. There's not a single drop of precipitation, no wind, even some sun, and Donna organises a mellow field trip focusing on one of this season's succeeders, giant petrel chicks, and one of the non-succeeders, kelp gulls.

Kelp gulls swallow limpets whole, removing the meat in their stom-achs, where the emptied shells are stacked one inside the other like hats on a hat stall, then regurgitated up in efficient packages. So all the shiny pale gold and dark silver limpet shells I see on the islands lodged in hollows, tucked into crevices, have been polished clean in the laundry of the kelp gulls' gut.

After the 1989 *Bahia Paraiso* oil spill, Bill installed eight limpet traps on islands. 'Limpets are very vulnerable to pollution and I wanted to see the impact. Traps would be a good way to do a check on recovery. It turned out as I thought. Fewer limpets.' But he hasn't written it up. 'No. Limpets live for 100 to 200 years, so why bother about ten?'

We find one of Bill's traps sited below a tall rock, on Laggard Island, where a pair of kelp gulls regurgitate. To Bill's annoyance the square of wire mesh covering the shallow box is covered in moss. This year the kelp gull pair began nest-building on top of the mesh but didn't continue, or breed. We pull away the moss to reach the shells beneath, putting them in a large, numbered, named ziploc bag for lab work early next season. Measuring size, age and changes in abundance will give an index of what the gulls ate. It feels very strange to be handling soft growing plants, destroying them even stranger.

The end of season banding of all giant petrel chicks has begun. We climb up to an anvil-shaped headland via a steep snow bank, kicking toeholds in as we climb. On the top giant petrel chicks sit alone in their nests. Donna: 'Their comfort level is increased if you stay from the wings forward.' Bill throws a piece of white bedsheet over a chick, which growls a low agitated, protesting *u u u urr*. Kneeling on the right, Donna feels for the leg and attaches a band (with Parmelee's pliers), the number is entered into the field notebook with nest number taken from a small metal tag discreetly located on adjacent rock, and Bill removes the cloth so the head is uncovered last, to minimise stress. Skua chicks get temporary bands because they wander from their nests. But giant petrel chicks do not move, so they aren't banded until the end of the season. If both parents are already banded, the chick is immediately built into Donna's long-term demographic study. Donna hands Heidi the working necklace, a long piece of string with light aluminium bands. Today she's being taught how to band geep chicks, a daunting responsibility, speed and accuracy required. Careful, conscientious Heidi tells me she needs to go over every procedure in her head each time. 'It doesn't come naturally.'

Headland finished, we slide back down the snow bank and cross a curve of ocean-facing beach. Whale bones from a minke that died in the 1970s, splintering and bleached, lie on the grey sand, and two fur seals' shoulder blades like small white dishes. Donna pounces on plastic: a white pot, a red film case and a bottle. Yesterday she found broken glass, the day before a nail. All human-related debris discovered washed up on the islands is collected and reported to CCAMLR. Bill checks, cleans and resets a second limpet trap, Donna and Heidi work the giant petrels in a difficult-to-access place, and I'm given thirty blissful minutes to myself. A spaced line of giant petrels are silhouetted against the sky among drifts of snow and whiskers of yellow lichen. Profiles of this island world: nothing intruding, nothing higher than a bird's head.

We eat lunch on an islet, backs against the rock wall of a small granite half-amphitheatre, gazing out at distant snow-packed, ice-hung mountains. We sit just a few steps up from the water, a gift from the calm seas. Bill says it's like an April day a month early. Today's downtime is long overdue, given the vile summer. The sea's surface is lightly overlain with miniature pyramids, lopsided squares and mimic island shapes, final remnants of ice floe and glacier. They pierce the intricate skeins of sea colour – two blues and cloud grey from the sky, and white reflected from the glacier. Everywhere there's the sound of air escaping, pops and small crickle-crackles, whisperings and fizzing – the trapped air of past lives of the world, quick little high sounds, flicking and cracking like crabs clicking their claws on a tropical beach. Donna notices something moving in the water and Bill says yes, a fast-flowing current runs past our islet's front steps.

So much knowledge of the area is wound and bound up in Bill. Showing him the first draft of a chapter he said he'd taken everything I'd described for granted. Was it how he felt, what he was thinking? Yes. Exactly. But it was coming through my eyes. In part that's unavoidable, and Bill is notoriously private. Meantime my outsider eyes and ears work overtime. I don't necessarily understand what scientists take for granted. I have to learn their verbal shorthand. Increasingly, I'm discovering the ways in which recording and interpreting today and yesterday is writing history; and comprehending the importance of the disciplines of history in understanding climate change. Records must be subjected to scrutiny, sources cross-checked, sparse data, and data with arbitrary start points, acknowledged. The past is an essential context for the present. But accumulating and processing, and the desire to write, are in conflict.

Bill is pushing for a precise list of everything I want to know, which is not possible. In the humanities not knowing what one needs to know in advance is a given. Matisse, on painting: 'I work without a theory. I am conscious above all of the forces involved, and find myself driven forward by an idea that I can really only grasp bit by bit as it grows with the picture.' Bill tries to understand how my mind works. Reasonable enough reciprocity – I'm trying to understand his. My laptop is a Mac, an outsider to the station systems. Bill, it turns out, has an unrelenting hatred of Macs. I'm guessing it's because his rather idiosyncratic system doesn't intersect with them, so Macs waste his time. The science-humanities interface grinds around, challenge mixed with frustration. At Palmer I'm a lone trader. Of necessity I work long, unbalanced days. As a freelancer my time is my income. But at my desk I'm not experiencing outside. And not taking whole evenings off for sociable chatting is a mistake. The pleasure of today's moments of relaxing make me realise I'm tired. But so are we all. My shoulder is knotted up from hours with my computer at the wrong height. Bill has a painful heel, plantar fascitis, not talked about. Neither of us is under fifty. In my case, under sixty. But Donna is in her thirties, Heidi in her twenties. It makes a difference.

After lunch we boat to Limitrophe, and Bill tramps off in his own space to collect the contents of three limpet traps. We wait in a pretty inlet with small pools, trickling water, starfish and seaweed; we are by the seaside, anywhere. Donna: 'There's nothing here.' Which means no data-achieving. Bill comes back with the news that all the traps are empty. He looks in the rock pools to find me some limpets – they can grow 60 millimetres across but he can't find any. Donna is watching timing. Today is 1 March, winter schedules have begun and the islands are officially open, with visiting restrictions lifted. The All Hands meeting is this afternoon, a day early to enable Bill to talk to Station members.

Back at Station Bill describes the kelp gulls' vulnerability to disturbance. A whole season's chicks can be lost, if someone walks among them. Where do they nest? All over, says Bill. South polar skua chicks, if disturbed, could be abandoned and die. Keep away from them, Bill advises. They hatched a month late and survivors are generally small and difficult to see. Use caution and care. In truth, I think, kelps, skuas and giant petrels have staked out most spaces.

Meeting over, we head out to the north islands, skirting loose jigsaws of brash, angular flotsam interspersed with large lumps of ice with small pieces stuck all over, like flies caught on flypaper. Beyond old Base N we turn along the inner flank of Norsel Point and enter Loudwater Cove, named for the frequency of its ice falls. Bill says the rapidly retreating glacier will meet up around the back, and Norsel Point will turn into an island. As we arrive a detached ice tower collapses into the sea. And as Heidi's giant petrel chick banding training continues up the slope of a hill there's another big ice fall – a sudden sharp slam like a door banging, with harsh close rumbles as if massive pieces of furniture are being shifted. It's so near that Bill worries about the impact wave damaging our zodiac.

Limpet shells at Norsel are thick as gravel, and at last I see a kelp gull's nest: a glorious green achievement, a metre-high cone. The average Adélie pebble pile would only come a quarter of the way up the side of this featherbed of beak-transported moss. Kelps tidy up their nests in autumn, then abandon them until they carry out moss replenishments in spring. The gulls stay in the vicinity of their breeding area all winter long, each pair with its own foraging space. Bill checks his traps and finds to his delight the largest haul of limpet shells yet, despite the gulls' failure to breed this season. 'More limpets than I've seen since 1994. I'm amazed by the sheer biomass these birds are moving from the intertidal zone.' Pointing to a gull: 'I banded that one in the 1970s.' Heidi: 'That's before I was born.'

There's a deflated zodiac by the boat ramp, vulnerable, floppy. Donna: 'Did you run over something, Jeff?' Then, 'It's one of ours!' Small rips and teeth marks show where a leopard has punctured the cone through the chew-preventers to the pontoon beneath. On the dock a fat crab-eater is lying sleeping. Bill: 'Everyone always claims any hauled-out animal is ill. It isn't. This animal is healthy.'

People have already been out on the islands, Hermit heading the list because it has the highest hill to climb. Donna gets from her box of magic tricks a small radio-controlled white boat. The water is so smooth that the boat skims around the inlet and a skua immediately swoops over, turns, makes several runs and then decides against the attack. Donna gets Bill driving, then laughing, relaxing. The evening, still calm, turns into a Station together event, with a return to past things like camp games. How many marshmallows can be crammed in the mouth at once while still saying an agreed word? Record: eighteen. We sit in a semicircle around a 'barbecue', an old steel tank converted into a dish on legs, burning small bits of waste wood, on the gravel, to avoid any risk of fire.

On Saturday a humpback swims slowly past our buildings, so clear we can see its entire length under the water, hear the breathing. Bill says whales often come in March because often there's krill. And on cue Stephanie excitedly tells me that she and Michael have caught their first krill in a month. They show it to me in a bottle, tiny, with a scrap of phytoplankton meal inside. And they caught a tunicate, a sea squirt 6 to 9 metres long, like a giant yellow-brown slimy worm. On Sunday, acoustic sampling near Hermit Island, their radio yelps, 'Humpbacks are breaching right out of the water.' There's an exodus from rooms and labs to look towards the islands. 'Holy cow! they are right next to us!' And an exodus to the zodiacs for whale-watching. In the night such a massive section of ice face crashes off the glacier that it registers on the seismic chart in T5; and Bio shakes.

Station talk, for support people, is often about each other, the insider gossip of in-groups, of *habitués*. Who knows whom. Who knows what about whom. Ice experience is what counts, with length of time 'on the ice' – which can be years working in the town of McMurdo – giving priority in room allocation. And the talk is about the buildings, what works and doesn't work, things that have gone wrong, or that aren't done the way the speaker knows to be better. The three Stations, the two seasons, the jobs. All the pleasures and pressures of working in Antarctica, then being catapulted out, maybe to the hub at Denver or perhaps to a vehicle, all there is, drive it out of storage with everything they own in it. Then flowing back through the various networks, back to the ice.

Scientists' talk is about other scientists. The in-worlds are smaller, the boundaries precise, the stakes higher: funding, promotion, reputations, positions on programmes, inclusion and exclusion, priority and protection of ideas, how data are accessed and used. Ambition and bruised egos, vulnerability and inadequacy. There's a brutality: too eager and ambitious a statement, too unhedged, unprotected, and the attack is swift. Generalist or premature statements are like late chicks: taken out. Mocked, dismissed. Caution can become paranoiac. It seems to me that science is not very different from everything else, full of uncertainties, hypotheses, false paths – exciting, dynamic. Scientists' responses to one another can warp the outside world's perception of how science is done, limiting understanding of and sympathy for the reality.

But overriding everything, in this small Station, is the essential lack of privacy, the gossip of any isolated community. People add themselves uninvited to any conversation, and listen. Two people walking together can trigger assumptions. Any talk can be overheard. Verbal jokes referencing other conversations can be taken out of context and misinterpreted. Sentences ideally should belong to the moment only. All possibilities in unguarded words are possible. Emotions should be kept contained, but the constraining creates emotion. Any row or argument can reverberate like a bell inside a tower. You must see every day the person who has offended you. You must see the person every day you have offended. The slightest word, or gesture, or lack of a word or gesture, can accrete meaning, even if devoid of intent. If intended, the intention hits harder, because there's nowhere to go. Petty acts of exclusion swell in confined spaces. Table seating can be devious, meal times minefields. One weekend there's a 'plate' dinner. There's no obvious way of knowing that writing your name on the small whiteboard in the galley means inclusion. So, when those who knew, and had put their names down, were sitting in the galley at tables arranged communally with white tablecloths having a special dinner, those who did not know stumbled in and left, fast, embarrassed. They hadn't been intentionally excluded. But it wasn't thought important to tell, or to explain.

My English husband went to boarding-school from the age of five. Communal living is deep in his soul: shared cold baths, loos with no doors, sleeping in dormitories. Surviving. Back in London I describe living on an Antarctic station and men laugh in an insider kind of way. Academics nod in affirmation of their competitive lives. An archaeologist detects similarities to the closed stressful environment of a dig. 'You either partner the woman you're working with, or you end up not speaking, enemies,' he says. Geologists recognise the symptoms of hard days in difficult places, cooped up with the same people. Archaeologists and geologists take the competitiveness for granted. But these insights come after I'm back. And writing is a solitary profession.

Small Antarctic bases are small communities. There is the potential pleasure – and pressure – of being a tight-knit group. Palmer this season has been a Camelot experience for several first-timers. In the Raytheon world Palmer is where you want to be, or the opposite: close-knit community against oil boom town like McMurdo, cosy instead of hardcore. Or it's claustrophobia versus possibilities, a joke, too small, too insular; up to fifteen hundred people live at McMurdo in the summer, so there's no hope of meeting everyone in a season.

But Antarctic communities can stifle. They breed kindness, and callousness. Sharing and competitiveness. Exposure to them coarsens some individuals, sensitises others. Jennifer described to me life in a small Alaskan community. People get embedded, she said. For her it was necessary to leave; the risk of spending too long is that you cease to remember how different the world outside is.

Bill says people are thrown out like a pot of dice onto the table at Gamage Point and have to get on. If they are no good, they get dealt with by the whole community. Frozen out. They have to behave or they are alone. They have to make a go of living together. Be sociable. And the amazing thing is, they do. 'That's why I like coming back; watching it happen.' But later he says: 'The doing of science is affected by what it's like to live here. Some scientists can't take it and leave. The pressures of being here are real. It's lonely, and public. To survive here, you need to understand the system.'

Which, with exquisite precision, he and Donna do.

27

A pair of legs, a pile of bones

Wednesday 6 to Sunday 10 March

Walking yesterday in the Back Yard, I watched a house-sized iceberg
bobbing and swaying in Arthur Harbour and then, with no apparent
outward sign, go beyond balance and turn upside down, water-carved
cellar becoming air-exposed attic. It took a lot more sea space than I'd
reckoned on to haul up and turn over the four-fifths of hidden ice from
below. An analogy of abrupt change – and of our ignorance. Much of
what happens in Antarctica is under the surface. We just see the bits
that show above.

At first I thought of icebergs as portraits of the physical world
– valleys and ridges, precipices, peaks, plains. Then I began to see
them as random disengaged slabs of their origins, ice sheet, ice shelves,
glacier tongue, split faces revealing the wedges of crevasses, the compressed layers of annual snowfalls, evidence of their creation. I began
to interpret their journeys: sea-smoothed undersides presented to the
sky; or bergs in partial tip, once-level tops tilted sharply to the ocean,
ski-slopes to oblivion. And I began to see the ice beneath the beauty,
the hardness beneath the seduction of form. Such ice: rock-hard, chisel,
steel-hard. Unyielding. Flint and obsidian. Solid. Once I passed along
the length of a tabular iceberg, frosted blue bulk, vertical walls rising
high above our ship. We had glanced the tip with a jarring jolt and
now, bruised bow, no alternative, were running the gauntlet of its
below-water unpredictability. Icebergs are like Antarctica. Care-less.
Or like the ocean, and space. They are, and continue to be, whether we
are here, notice, die, live, are sad or happy. Icebergs exist, until they
cease to exist.

A wall of boxes hems me in at my work-bench. Equipment from the labs is being piled into the Dive Locker, and my days here are numbered. But through the Field Room door I can still see a small square of Antarctica. A leopard seal swims past up Hero Inlet, big head and shoulders above the surface, then under again. Invisible. But later I hear a strange noise, like a woman hurt: a quick, loud cry of pain. An Adélie is standing, feet wide apart, on a snowy rock, just outside. It looks thin, and its black back feathers seem greyish, with a few white feathers adrift in its smooth white chest. It moves its head from side to side, its flippers up and down, up and down. I go back to work, and the sound becomes a sighing honk. At lunch Bill explains it's lonely. When penguins are about to start moulting, they come ashore. They don't want to be alone for the two weeks of moult, so they call. The hectic commitment and noise of summer are over. The pace now is entirely different, just single birds or small groups, quiet. Once their new feathers are in place, they head out to sea to feed, then tend to return in April, poke around the nest sites, perhaps do a bit of work. By May they've generally gone.

Bill talks through this complex season, cautious, guarded, tracking the factors. 'The problem is that climate change has always been used generically to explain a lot of things. It's an easy catchphrase for everything, but it can't explain something specifically. Climate change may account for an effect. But it needs data. And it needs an understanding of the mechanisms by which it is operating. What I have done is to propose mechanisms by which climate change is operating, in the sea ice, and via snow on the landscape. Snow and ice together are creating havoc for species that are truly polar.' By which Bill means less ice, and more snow. Less ice causing havoc to a polar species is not surprising, but more snow seems, as usual, counter-intuitive. Rising temperatures are bringing increasing precipitation, and more snow at Palmer is where the specifics of locality kick in, the specifics of local landscape.

Last time at Palmer, Bill told me about plans to build experimental snow barriers to observe the effects on nesting Adélies. The barriers are not yet built. I don't understand why barriers are necessary. Surely,

I say, nature has already done it. We've just seen the 'natural experiment' and the results, on Litchfield. But no, Bill says, it's necessary to understand the mechanism more precisely, exactly how it works. Plus there's politics. 'Ecologists reviewing our work are clamouring for it to be more experimental.'

Bill: The opportunity, and challenge, is making connections from the particular to the wider environment. We have empirical data. Now we have an intellectual basis for taking territorial and marine ecology together in Antarctica. It's clear that to really understand we need both, territorial and marine – play them back and forth. Take the south polar skuas as an example. This year, delayed by the snow, the skuas bred a month late. 'By doing the same things year after year we are able to distinguish an anomalous year. We know that late breeding isn't the birds. It's the environment. This year eggs were laid in a quick window of time. There were fish around to induce breeding, and the breeding season showed extreme synchrony. But late breeding has consequences, and our work has been to record them. Most nests only had single eggs, most eggs were light in weight, and birds that had two chicks lost one immediately. During anomalous years horrific numbers of chicks disappear. It's really difficult to know what takes a chick out. It might be weaker and so less able to hide. Parents might have to search longer for food. In a good food year more time is spent guarding the nest. Neighbours might be hungry, so prey on next-door eggs or chicks. It has to be speculation how a chick disappears. We don't know. But always we come back to the mechanisms; that's what we are trying to get at. You could say that a weak chick is more vulnerable. But,' and Bill pushes in another variable, 'you could say a chick is more vulnerable because of our work.'

With south polar skuas, reproductive success rate and the growth rate of chicks are both clearly related to the availability of fish. 'But this year the problem is not whether there are fish. Because the timing of breeding was affected by snow, there was a mismatch between time of laying and the availability of fish. The peak of fish was not met by the peak of need. Always keep in mind the difference between abundance and availability. Food could be very abundant. But major storms have been coming through every three days, affecting availability. Water is stirred up, it's rough; the food is there, but you can't get it. But we can determine from this season that breeding late because of the snow has impacted on the availability of food. By monitoring south polar diets – and Adélie diets – some idea is gained of what is happening to the fish population. Almost everything we do aids in understanding the structure and variability of the food webs here. For south polar skuas, reproductive success is either boom or bust. Fast, or forage. The final results are not yet in. But this year is probably going to be bust.'

That 'probably' seems an extreme of caution. A week into March, Bill is prepared to offer some other summaries of the season. Blue-eyed shags have had a normal year, although final figures are not yet in. But kelp gulls are a complete breeding failure. Last season 120 pairs of kelps were censused fledging 1.2 to 1.3 chicks per pair: a bit on the low side, but successful. This season is catastrophic.

Far to the south, near McMurdo, the vast Adélie colonies at Cape Bird, Cape Crozier and the most southerly of all, Cape Royds, where Shackleton sited his hut in 1907, are suffering heavy losses. As are the emperor penguin colonies. Massive icebergs adrift from the Ross Ice Shelf, extensive fast ice and multi-year sea ice, are blocking breeding pairs from reaching their traditional nest sites and forcing birds into unmanageable journeys to and from their feeding grounds. This has been US Adélie penguin specialist David Ainley's pitch since the 1960s. Ainley sends me descriptions of fierce katabatic winds blowing eggs and chicks out of nests, of buried birds, starving birds, massive failures to reproduce. 'Hardcore and persistent ... it seems that everything that could go wrong has gone wrong for Adélie penguins this year in the southern Ross Sea.' Bill: 'We are two very strong-willed people. He enjoys and does super-large-scale thinking. My focus is here. We are doing similar work, but his emphasis is heavily on foraging.'

The birds at McMurdo operate in a polar system only 1,400 kilometres from the south pole. They are having almost identical problems to Palmer's birds, but with different causes. Bill: 'Both stories, there and here, show the ecological and physiological limits that these birds can deal with. But the consequences are the same. Some threshold has been exceeded by the environment which they can't handle.'

Heidi and Donna arrive back from the latest south polar skua rounds on Shortcut. Donna shows me five centimetres of soft, whitish shaft like the handle of a feather duster, with three and a half centimetres of black feather poking out and a puff of down at the very end. It's a primary feather. It belonged to a chick from F ridge, one of this season's potential success stories. The chick was first recorded on 3 February, among the earliest to hatch, the late season finally under way. A first chick with no siblings, it was doing well. Its primaries were growing to order. Then today – nothing. Or, to be precise, a pile of bones. A similar well-developed chick from E ridge weighed 310 grams on 14 February, 514 on the 23rd, 790 by the 28th. Today it is a pair of legs, with the tape attached. Both E and F ridge chicks, Donna worked out, were 36–37 days old when they died. They'd had a better chance of making it through this dismal season than many.

There have been plenty of eggs that didn't hatch, or losses with younger chicks. No carcasses. But today Donna and Heidi report losses, losses, with chicks lying, partially munched, at their nest sites. Heidi says she saw some fluff fluttering and found a cowering chick with a south polar skua standing in front, wings at full stretch. They grabbed the chick and took it back to its territory. One 'rebirth', a small chick sitting under rocks that had not been seen and so was thought to have died, was also found. Heidi: 'A carcass has no potential for rebirth. That's what's so disappointing.'

The short run of calm, dry days breaks, and it's back to rain, snow and wind. Most of the scientists leave in less than a week; the season is almost over. Rebecca says they've had to compromise, adjust, improvise to get the persistent organic pollutants samples completed; it may not be optimal, but the work must be achieved. Air is the main transport factor, the key to everything. But POPs are most diluted in air, and air is the hardest to measure. She has a hundred filter papers, each representing 1,000 to 1,800 cubic metres of air sampled over 24–36 hours, all stored in the freezer ready for shipment, next to the sea water samples – each a bathtubful, filtered for ten hours, reduced to a filter paper stained brownish green in a clean glass jar, plus filtered in a carbon column, capped off. All the POPs Antarctic field work will end up in the States. Large quantities of raw Palmer product reduced to tiny ultra-clean samples for very high-tech, sensitive analysis identifying the chemicals, seeing their signature, measuring. The Arctic has an international treaty to phase out certain chemicals. Rebecca's desire is to develop a scientific base of information that could benefit Antarctica. The work, she says, could take her the next ten to twenty years down here.

Two of this season's young researchers won't return as scientists. Both want to be, but they say they've had to prove it by doing tedious lab work, not requiring their level of qualifications, for little pay and long hours. They are coming back as support staff, earning twice as much – and, they laugh, getting paid for doing fifteen minutes stretching every morning. Raytheon's gain, science's loss.

T5's owner gives me an introduction to the crowded machines of the six science projects requiring his routine, on-time, accurate attention – from listening to the signals generated by lightning to monitoring air for nuclear fall-out, detecting delicate earth motions, to monitoring UV. 'Diligence is difficult.' I get tours of the power centres driving the station: fuel and generators. 'There's no locks on the pump house door, the valves on tanks. Here you don't even think about it.' Wendy shows me around the kitchens, up to two years' food tracked on a daily basis, stored at three temperatures: freezer, refrigerator and room, although that's too warm, and bugs imported south from Chile can survive. I find out the two fates of sea water pumped in from 20 metres along the seabed to GWR where it divides. Some stays as sea water to be used in the toilets, and travels on to the macerator in the blue hut by Gamage Point. All waste matter on Station from faeces to food ends up in the macerator which works non-stop, de-lumping what it receives to a fine soup, then shoving it out into the sea. The rest of the sea water is processed by reverse osmosis into fresh water for showers, laundry, drinking,

cooking, the lab and the fire tank. 'We make what we use' – 50.4 US
gallons a day for each of us.

On Saturday 9 March our contribution to Earth Day is a clean-up
outing to remove rubbish from the first US base, Old Palmer, and Base
N. The weather is vile, with continuous rain ensuring slippery rocks,
and a 30-knot wind heaping up rough seas. We arrive wet and get
wetter working. Our bag lunches stay in our bags. The abrupt hills of
Litchfield's sombre backside lie moored across the seaward view, partly
obscured by Humble. Our promontory with the Station buildings, the
openness of their setting, the small-scale grandeur of Arthur Harbour,
is completely hidden. The wind gusts harshly down the valley off the
ice sheet behind. In 1955, when Base N opened, the frozen slope up to
the piedmont began at the hut's back door. Now the ice starts out of
sight, fifteen minutes' walk away: further proof of warming. In 1955
no fur seals were here. Now they occupy the site, swimming in the cove
while we toil. Debris from the site was cleared by the US Antarctic
Program in the 1990–91 season. There's not much evidence, apart from
sections of concrete base and assorted foundations, for human occupation. But I do find small lumps of melted glass, presumably evidence of
the fire which consumed the Base N hut in two hours on the morning
of 28 December 1971, when two visiting British expeditioners using a
blow-torch to remove paint accidentally set fire to old packing between
ceiling and loft floor. The fire burned a building with a chequered
history of occupation after 1958: Palmer biologists using it for a lab
and relaxation, plus brief British visits and clean-ups.

But fourteen of us do pick up enough rusting nails, lengths of wire,
bits of wood, minor pieces of equipment, broken china and a couple
of identifiable objects – a spoon, a knife, a hinge – to fill four sacks.
There's a Palmer safety cache on the site, and replacement blue barrels
are lugged up the rocks and placed in position among fur seal faeces.

After less than two hours, soaked to the skin, gloves and boots filthy,
we withdraw and go back to our comfortable galley in our comfortable
warm buildings to eat our picnic lunches by the contained flames in
our little cast-iron stove. Heidi, Bill and Donna have been at Norsel

removing wooden stakes, stainless-steel stakes and welding wire –
rubbish left by a previous researcher.

In the evening snow begins, from the south. It's still falling on
Sunday morning. Jennifer, on her morning gaze, says the snow is wet
and sticking. When she arrived at Palmer, she pulled down a picture
from the slide show in her brain of the small Eskimo village of Wales
where she had worked in 'bush' Alaska, on the edge of the water – wet
maritime – and found it similar. The only difference is what grows.
She finds being at Palmer familiar, and takes the weather for granted.
She's used to living in small communities, and being here is comfortable and easy.

The US Antarctic Program includes people who've come from
extreme cold and sub-polar conditions. Experience does qualify judgement. One of the geologists on the *Gould*, talking about the peninsula:
'I hate the way writers say "I was cold. It was a gale. I was frostbitten."
This is the way it is, describing it isn't interesting. And this isn't really
cold. This isn't really a gale. Minnesota: that's cold. Sixty below. This
is an extreme version of northern Scotland. People could live here. They
do in Siberia.'

He's right, of course.

28

Last island on the geep chick banding tour

Monday 11 to Friday 15 March

The single chicks look almost overgrown: teenage schoolboys in short trousers. But still they sit, anchored to the mooring of their shallow soup bowls of nests, splashes of excreta radiating out across the stones from one side. Their fine white down is now wispy, loose-weave, with a sense of smooth wing feathers, darker colour beneath. Occasional adults watch in attendance near by. Here on Litchfield the Adélie colonies have been abandoned, the brown skua pairs have failed to make a living, but the giant petrels are thriving. We work our way up a steep hill overlooking the sea, facing north into the prevailing wind. Prime giant petrel territory. Silver-grey ledges and narrow benches of flat paving stones, or leaning headstones, with a skim of grass and moss between, provide ideal nest sites. Yesterday's snow coats the rocks, the temperature has not risen above freezing, and surfaces are very slippery. But the massive task of banding all Palmer's giant petrel chicks has still to be completed.

Donna holds the pliers in her right hand with a band already in place, the next four ready to go on the fingers of her left hand. Heidi sheets the first chick, holding the cloth in place either side of the body, and Donna is on her knees, feeling for the chick's right leg. A quick squeeze, the band is on and she's calling nest number and band number for Bill to note. Four to five seconds. Heidi pulls the white sheet off from the front, and we zigzag up to the next nest. Donna navigates from the map in her mind, moving fast, not pausing. The nest numbers are already programmed in her memory, the route planned in advance, time taken to get around and finish calculated. She counts everything, wherever she goes. Including stairs.

Nest 29 has a chick with subtle grey-brown feathers showing through ragged, flaky fluff. What gender? Donna guesses female, but size affects gender, which is affected by age. The adult in attendance was banded as a chick six years ago. Both parents nasty, comments Donna. Three nests further on a chick does a mini-spit, caught mainly in the sheet though a bit hits Heidi's trousers. Donna: 'A chick takes cues to be calm from the parents.' And these are bad-tempered. The smelly piece of bedsheet is bundled away and replaced by another.

The hill opens upwards on to an easy summit with views down to the pebbly sweep of abandoned Adélie colonies, across to the mini-mountains, around to snow banks that have lasted all summer, to the beaches and pieces of sea ice frozen to the shore. Giant petrels circle in the grey sky preparing to land: feet forward, head out and up, like 747s. I think about the accumulation of days and nights a chick occupies the single small space of its pebble nest. The one egg hatched during December's long light, the chick, unable to thermo-regulate, protected in the brood pouch during January's warmth. February, growing visibly but still down-covered. In mid-March, with new-fledged feathers, the chick might take its first small, slow steps, perhaps a metre from the nest. Try flapping its wings. Get back to the nest. Try walking with wings flapping, a few metres, even ten. In another month – or two (males take longer to raise) – it will attempt a stumbling, swaying run, launch briefly into the wind, land. Walk back to the nest. Lift off into a gust and spill back down, manage for longer, get dumped. Again. And again. Until on cold, clear, windy days in early winter, hungry, no longer being fed, this season's fledglings will leave their nests for good and fly away.

Ever the realist, short of funds, Donna uses aluminium bands provided free. Giant petrels tuck their legs up in flight, and the light-weight bands are not an aerodynamic inconvenience. But cheap bands have a finite lifespan. Birds smack into rocks, they fight, and chicks are risky recipients: their survival rate is low. For adults, the proven breeders, she uses her own preferred bands, paid for out of her own funds.

Chick banding takes longer than calculated, so plans change. Donna and Heidi will band in the 'high country', while Bill and I do the seal loop, counting all the elephant, fur and Weddell seals – everything in the water as well as on land, because animals may well have been scared into the sea. The mammal census started with Parmelee in the 1970s, and the sequence is building. We walk a clockwise route beginning on a wide apron upholstered in greeny-gold moss the colour of 1970s' carpet, worn, threadbare, flattened and reduced by seals' bodies and their effluent. The moss is frozen; the ground beneath is frozen. A group of elephant seals have taken up a position on top of a hillock on once pristine moss beds, but the rest are tight-packed in the centre of the apron, flippers over each other, steam rising from their bodies, scratching. They sneeze, and the length of their bodies quivers and jerks. Their moulting pelts are peeling off in big sections like sheets of sunburned skin, curling up at the edges. They gaze across at us with brown, unblinking eyes, roar a bit, grumble but don't adjust their positions. The fur seals are scattered beyond like a fling of dice, dark brown bodies across the moss, with squiggles of deep scarlet-red faeces. They rear up on their front flippers, engage one to one, back off, circle, endlessly posturing: old heavy bulls with thick necks; young males supple, agile. Bill minds them less than elephant seals. They don't damage the moss as much, arrive later than the Adélies, don't colonise the nest sites and leave in June or July. But furs aren't easy to predict. They can run faster than a human, and bite – like big dogs, according to Donna, who admires them. They look us in the eye, keeping the eye line by moving their heads and upper bodies. We skirt them as far as possible, but several decide on attack, making their rush from behind after we've passed. Bill grabs my trusty Swiss mountain stick and deflects them. Then he finds a bamboo, which is better because it's longer.

Walking past the abandoned Adélie nest sites, we do, after all, see penguins. One, thin with moult-induced starvation, still needs a week to grow new tail feathers. Another is part-way through the moult, chest feathers like thistle-down; the third is just beginning, fat, fluffed out and scruffy. They stand in a line on top of a south-facing snow bank, where thousands of Adélies have stood to moult over the years. Curious

black spines stick out of the icy snow like sparse hedgehogs: genera-
tions of tail feathers, unexpected remains layering the ground. Evidence
of the process of living.

The secret valley hung with moss cascades, so beautiful two months
ago, is now staked out by fur seals. Snow covers the moss, stripping
out colour and texture. The stream is silent, frozen. We walk through
quickly, focusing on the fur seals' intentions, to find Donna and Heidi
waiting by the wide cove on the other side. The sky is the uniform grey
of impending snow; it's getting late; Donna and Heidi are tired. But
Bill wants to finish the work. Litchfield is a Specially Protected Area,
human impact must be kept to a minimum and whatever is needed
done, if possible, on one visit. So we set off for the 'back country' in
a silent line under the cliffs, scrambling over snow banks, boulders
and lumps of tossed-up sea ice, waves washing high up the beach, to a
narrow valley, its entrance guarded by fur seals. The route to the giant
petrels nesting here is up a frozen slope of last winter's snow, packed
into a ravine. Donna tackles it on her hands and knees.

I wait in the valley, against the valley wall, standing still. The posse
of fur seals watch me, yelping their short *urrhh. Urrhh. Urrhh.* There's
the sound of surf, in the distance, and the cries of south polar skuas. The
heavy old bull rotates to the edge of his group facing across, *HRUFF.
Hruff! Hruff!* Stares. Then rotates back into the middle. Seven younger
seals flatten the moss in casual rounds of neck-rearing and shoving.
Litchfield has style. It has enticing, satisfying proportions, large-scale
landscapes reduced to an encompassable, manageable scale. It's possible
to walk, climb, explore. This valley, slicing briefly through the back
country, is like a small canyon, floor planted with individual boulders
placed amid the moss, as though in a Japanese temple garden. Moss
grows on every tiny half-inch ledge in the cliff faces, scribbling the
rocks with a calligraphy of green lines. Yellow lichens define the verti-
cals. Opposite, the crags hold a miniature alpine meadow, moss-filled,
high and north-facing, to catch the sun.

Snow begins to fall, fine crystals outlining the contours of the dark
rocks. The snow sticks to the moss on each tiny ledge, and the green

scribble turns white. I watch the landscape transforming and think, nothing can appear – could ever appear – however long I stand here. Nothing lurks in the hillside behind. There are only birds, their chicks and seals. No small animal will scuttle between the rocks or burrow by my feet. No beetles or ants will navigate the surfaces. If I choose to sit on a rock, I have it to myself, entirely. No insect will fly through the air – buzz me, or bumble by or soar on the winds. None. Ever. Only microscopic, slow-moving nematodes, tardigrades and rotifers, deep in the moss, sheltering under pebbles. Only parasites – fleas, ticks, lice, mites, worms – burrowing, hiding, clinging on to their bird and seal hosts. The extraordinary release of Antarctica. The emptiness.

Sometimes when I walk over the grey piled pebble-rocks of an Antarctic beach, I think of Tasmania to the north, but not so very far north, and being warned not to go on the blue-grey pebbles of a certain beach because poisonous snakes lived among them. The shock of finding that beach pebbles could be dangerous. But here, in Antarctica, the danger is the inanimate rock: unstable, poised, en route, not bedded down by roots, wedged by earth or sand. Or – once the temperature is minus celcius – seashore rocks lethally slippery with an invisible coating of salt ice. Or scattered pieces of ice, piled shards or rounded boulder ice smoothed by the sea, but ice nevertheless, washed up on the tides and covered, as is happening now, by a layer of fine-grained snow that is also coating the rocks the ice sits among: all lumps equally white-dusted but not equally viable to walk on.

The skuas begin to mewl and scold, so I know Bill, Donna and Heidi are on their way back. Bill has a certain mournfulness of mood. Everyone is tired.

The valley dips ahead towards a bay facing out to open ocean and the distant Joubin Islands. By the seashore rocks Bill picks up a small ball – just a head, with the first lappet of skin and feathers below, but enough to see the thin black line across the white, the signature of a chinstrap penguin. Heidi finds a newly dead south polar skua, still soft and pliable. Its head droops forward then rolls back. Its keel bone is starkly prominent. Bill says it has a broken wing and has died of starvation. I hold its light body, look at the distinguishing band of white feathers on the wings, the way the bird can sit down on the foot and the part of the leg below the first knobbly joint. I ask to see where the measurements are made on the culmen – the bill. No angry, pinioned bird, but time to examine. Measurement starts from the point where the feathers begin and goes to the tip of the hook. The deepest point is from the 'notch' under the bill where the 'v' starts, up past the nostril. So now I know.

Heidi and Donna finish the remaining giant petrel chicks and we retrace our route, Donna holding a bottle-shaped foam float, Bill the south polar skua wrapped in the white sheet, a decent shroud. It will be frozen and taken back to the United States: a specimen in someone's research. At the landing place the rocks have a coating of white, and our zodiac is layered in snow. Donna's thermos of tea has turned into America's favourite, iced tea.

Litchfield giant petrel chicks have experienced few losses. Breeding pairs have achieved a success rate of 88 per cent, maintaining standard patterns for chick rearing. This year the overall breeding success rate for 464 breeding pairs of giant petrels at Palmer study sites is 85–90 per cent. Three hundred and sixty-six chicks have been banded, leaving just the study-site chicks at Humble to do. Since 1974 chick production has increased at 2.7 per cent per annum. Breeding populations in the vicinity of Palmer may have doubled over the last twenty years.

In many ways Donna's work is first-stage science: observing, collating. She has seven to eight years of chick growth and fledgling data, a massive concentrated effort. But 0–1 is the most vulnerable year for wildlife, and attaching expensive transmitters to fledglings is not currently realistic. Fledglings leave Palmer as winter starts, usually by 15 May, all by the 27th. Donna: 'They can depart fat – but if the wind hits, where do they end up? The huge influences of space and time, of being where, when. Giant petrels have circumpolar distribution. They could be 9,000 kilometres north by July. They must find food, manage Southern Ocean storms, live, gain strength.' Sometimes, in South Africa, New Zealand, Australia, people find young birds, washed up exhausted, on beaches. The journeys these birds make are mighty, and testing.

The years from 1–10, until the birds become adults, are the great unknown, the biggest gap in our knowledge. Donna: 'I want to know who feeds where, and what trouble they are getting into. The non-breeders, and their survival, regulate the breeding population.' The youngest non-breeder she has seen come back to the Palmer islands was aged four. Birds start reappearing to try courtship, stake claims, build play-nests. Males can breed from six, females at eight. Not all the survivors from the absent years return to their island, but so far there is no information of a Palmer bird breeding anywhere else. Ninety per cent are faithful to one site. And once they reach breeding age, giant petrels can live thirty or forty years, perhaps fifty – it isn't yet known. They make strong pair bonds. If one of a pair dies, the other takes time to build up another bond.

But giant petrels follow ships. Fisheries in the Southern Ocean, using near-surface baits trailed on long lines, take an ever increasing toll of seabirds, tens of thousands, annually. Legal fisheries using new fishing techniques report a rapid decline in numbers taken. But seabirds lost to illegal fisheries are significant, and incalculable. Albatrosses and giant petrels, as surface feeders, are especially vulnerable. Donna has found recent and rapid increases in long-line hook retrievals at giant petrel nests on the Palmer islands. Hooks can come back with the birds. But usually it is the birds that do not come back. And the newly-fledged chicks? Airborne for the first time in May, heading north, they will fly straight into the July tuna fishing, off the west coast of South America.

I've needed to stay at Palmer till the end of summer, the slide into autumn, to see the islands begin the turn towards winter. The signs and signals are all around in the physical world. From the top of the hill on Litchfield I'd seen the *Gould*, looking clumsily large, moored along our little dock. As we steer around its orange bulk at 6.30 p.m., the sky is lemon. The sun sets at 7.30 straight down into the sea. A new, shocking speed: the clocks have changed, and we've lost an hour. And the signals are in our imported human world too. The winterers have arrived. This is serious changeover time.

Next day confident strangers are everywhere: Raytheon old hands with ice experience, though not at Palmer. Only seven women, mostly men, wintering support staff and construction people here to dismantle and rebuild the labs. They roam and investigate, finding the rat runs, checking opportunities, occupying the spaces. We are in the way, to be tolerated, before the real business of winter gets going. I doubt they are knitters.

People walk into the Dive Locker. Conflict of interest flourishes. Who does a space belong to? Who is it for: Raytheon, or scientists? A staffer shows his replacement around, grumbling: NSF forces too much science on us. They should cut the programmes. If the labs need remodelling, they shouldn't have science being done at the same time. The old, constant Antarctic equation, science versus the enablers, tussles first played out on board nineteenth-century ships, now rumbling around the structures on land. Scientists, pressured, demanding, unpredictable, needing time, space, resources. Support expanding, becoming an end in itself; science having, especially here on American stations, to deal with civilian contractors, exacerbated by the tradition of military support, with rigid US Navy Department of Defense rules and schedules.

Surrounded by boxes of fragile laboratory equipment at my Dive Locker bench, I've become invisible. But these are my last moments here. Serious lab dismantling has already begun. My final Palmer work space will be the small shelf at the end of my bunk used universally as the step up to the top bunk but about to resume its status as a desk. Since Steffi left, I have had the privilege of a room to myself, and now it will double as a study.

As the sun sets towards a clear horizon, the word gets around that there may be a chance for a green flash. I've hunted the sight in the Australian desert; now the Antarctic desert delivers. We watch the reflection of the setting sun in the window, turn at the right moment and capture, small and fast, an arc of bright green above the yellow-gold ball.

At 8.30 p.m. the *Gould* returns from a day's science as Bob and the new Station manager Joe Pettit, compete for how long they can each keep four 'fireballs' (hot lollipops) in their mouths. Next morning, grey and raw, 70,000 US gallons of No. 2 Diesel treated to withstand gelling to −20°F is transferred from the *Gould*'s tanks, briefly over Antarctic soil, into our two storage tanks for winter use, using the ship's pumps and so their responsibility. Five hours of monitoring, each connection poised over white absorbing paper in a tray, manned by a guy on a chair. The rocks along our shore, licked by the sea, are coated in black ice. Vehicles grind past shifting cargo. The twelve leavers move on to the ship, becoming half-way people. They walk down the gangplank back on shore, but no longer to their own space. They go back up on the ship, but are not yet fully integrated.

Scientists on the *Gould* give me a glimpse into other intense commitments to unravelling climate change along the Antarctic Peninsula. Three years ago, as part of a project tracking climates over the last twelve thousand years, they removed a 45-metre-long core from the sea-bed in the basin of the Palmer Deep, south of Anvers. Ocean-bed sediments such as this are a key source of information, along with coral reefs and ice cores. Now another coring is being made of surface muds. The scientists are trying to find out what can be learned about the history of currents, about the important Antarctic deep bottom water, by comparing the two cores. Yesterday, 12 nautical miles south, a mooring measuring water temperature and salinity for the past three years was pulled up from 820 metres down in the Palmer Deep. The data was read off, the batteries replaced, the mooring serviced and lowered back down. Phytoplankton, significant creators of oxygen and absorbers of CO_2, are negatively affected by a really rapid drop in the temperature of water in winter. Ongoing studies of the ocean interact tightly with studies of the living things using the ocean. So much is going on, so much to find out, they say: the synthesis waits to be done. But plate tectonics to them are the definers of climate. When land masses change relationships to each other, sea currents change: and climate changes.

Thursday 14 March. Last night's party lasted till this morning. I noticed, after my drive to clean out my conversational exaggerations, that almost everybody now says 'absolutely' and 'marvellous'. A confused new winterer stumbled through my door in the early hours to announce he wanted to go home. It's his second premature pull-out from Antarctica.

The *Gould* will leave just after breakfast. Departures are uncomfortable. Some linger, saying emotional goodbyes; some march resolutely on. Tim, the Boy Scout, suffers the wrench of first-time abandoning Antarctica I remember so well. Antarctica can grab your soul, and you are committed. Heidi is finally going, long overdue for a break. Consistently kind and thoughtful: 'I don't think like a biologist. I don't naturally think, for example, "why has a skua starved?" I think, "poor skua". I trained in science, but I'm interested in people.' Recently she rescued me from very visible humiliation. I didn't understand how a computer programme worked: 'My father's a 747 pilot, and he doesn't either.' Unanswerable, and then she showed me how to access what I had long needed.

At the stern the Filipino crew cluster joyfully, waiting to see just one thing: the Antarctic plunge. Lumps of ice churn between ship and dock, so timing matters. It's never quite clear who will plunge. Some apparently fully clothed suddenly strip to swimsuits and dive in. Some whip off everything. The Station's legendary swimmer who, according to tradition, invented the plunge, to celebrate the departure of two science teams, lasts this time six minutes, wearing gloves, socks and sandals, floating, sitting, swimming, not expending energy. Everyone else is out in thirty seconds to two minutes, goose-pimpled, yelping, into the hot tub to revive. Ever-willing Jordan in a tie-dye top and nothing else, makes it to the zodiacs then can't get ashore. The ship disappears fast.

We summerers are down to the final ten. Standing huddled in my coat, I feel a bit like the last seals, not yet culled. In a fortnight we will be gone.

Station is very different. The new Raytheon people are vigorous, noisier, doors slamming, more smokers, breakfast full at 7.15 instead of empty. Already construction workers are swarming over half the labs, rooting out walls, tearing out fixtures – moving fast while things aren't frozen. In Bio our berths are stifling. My next-door neighbours show me how to stuff a towel over the floor heater and the temperature goes down 15°. Fire alarm systems are being thoroughly attended to, with every room checked and raucous alarms going off to make sure we can hear wherever we are – head under the pillow, body under the shower, part of Station Notification Devices and Circuit Testing, as big JB with the ponytail gleefully informs us, by email. And none of these left-over British terms any more. Bio's ground floor is now the first floor, as in all US buildings. The new labs will have interchangeable units, uniform: one reason why the Birder Box does not fit. The Dive Locker has some wooden drawers built by the same carpenter: individual, well made. In small, confined spaces examples of the human spirit give pleasure. Individuality, even quirkiness of design, one-off solutions, good work-manship, surfaces of wood, can all aid people's sanity and well-being. At least I think so. Yesterday a new guy snaffled one of the few indi-vidual mugs in the galley, my favourite. At the Scott Polar Research Institute in Cambridge we have our own mugs for communal morning and afternoon tea. On departure we put them in a kitchen cupboard in the hope of return. Last time mine with a first draft of Keats's 'Ode to a Nightingale' was lodged between mugs from a Polish Antarctic station and the Metropolitan Museum in New York. I'm told that at the new Palmer individual space is thought to be appropriate only in bedrooms; but they are shared, and waking time isn't spent there. I ask where the sewing machine will go in the remodelled space: it's no one's concern, so probably there will be no place. The sewing machine belongs to the Station ethos I've experienced here: the idea of make and mend, plus make and create. There's no dedicated space for creative activities. Perhaps that's one reason why the knitters annoyed the non-knitters. Balls of 'yarn' around, untidy.

Bill and Donna have less than two weeks to get through the remain-ing fieldwork plus indoor chores, and database assembly. Bill: 'I want to leave with this season's conclusions. I've always done this and try to keep up the pattern. The figures are put in in random order. Now I can access them.'

Bill is finally in the chief scientist's office on the first floor of Bio (except now, with JB's renaming, second floor). It's Sunday morning. St Patrick's Day, we are informed, but no parade. Bill and I are talking: bright upholstered chairs, a carpet, a proper desk. Through the small square of window the sky is a brilliant, unexpected, clear blue.

Suddenly Bill starts. 'I've never seen so much of Mount William from here before.'

On days of clarity, magnificent, snow-covered Mount William, 1,515 metres high, is visible from our small offshore islands. But never, according to Bill, could Mount William be seen from the chief scientist's desk here, in this office. Then – ten years ago, sitting at the desk – Bill saw the tip of the peak emerging above the line of the piedmont. Every year, as the ice sheet slumped, as its bulk reduced, more of the peak was revealed. Until suddenly, this rare blue-sky morning, like a child's drawing, a distant triangle of glittering white breaks the long ridge line of the ice; and the whole summit of Mount William as far down as the saddle is clearly visible, as well as a lower peak.

The ice is reducing on a major scale, shrinking in height, and losing mass from the front, retreating. Bill: 'Up on the glacier at the back of the Station, you can hear the sound of water rushing inside and below. The core temperature is being altered by warming. The severity of the issue is all-encompassing.'

Several hard, sharp cracks, then a massive roar. Blocks of ice tumble from the ice cliffs across the bay into the sea, rolling heavily. The impact waves come radiating across the harbour, hitting urgently, ungently, against our shore. Bill, thoughtfully: 'I'm doing all these studies of climate change. And, in fact, it is visible. I can see climate change from my window. I can sit at my desk and observe it. It's so staggering what has to happen, for this peak of Mount William to be revealed. Mount William is evidence.'

As I leave: 'You can record climate change from the women's head upstairs.'

Eleven years ago, if I had been at Palmer, I wouldn't have seen Mount William from the loo. But now, gazing through the small double-glazed window, I can. As long as the sky is blue, and the sun is shining: which, this summer, has hardly ever happened.

29

Summer's end

Monday 18 to Wednesday 27 March

Out on Humble the bedraggled Adélies have gone; no seals lie across the route; the brown skuas don't harass – one pair lost their chicks, the surviving chick of the other pair has fledged. The vigour and stress are over. Everything is dry, rocks, elephant seal swill area. Colours are more embracing in March's gentler light, smells minimal in autumn's colder air. On the giant petrel uplands non-breeders play at nest-building. The big chicks, now 4–6 kilograms, are skittish or grumpy. Bill has officially lost the horse bet, and Donna debates a paint mare or a nice sorrel. 'I had faith in my scavenging loves. There was a part of me that, despite early egg losses and apparently infertile layings, believed they'd pull off a decent season.'

On Station the salt ice layering the rocks licked by the tides does not go. Snow covers the ground in a distinct layer, and knowledgeable newcomers take small shuffling steps on the boardwalks, checking its stickability. The great snow bank near the hot tub that last spring was dug into for a cave, then became a bridge, is now a 4-metre-long knee-high ridge. But it's still there. This summer the snow bank beside the skuas' pool beyond T5 didn't melt either, and now the pool is frozen, snow surface dented with birds' footprints. When the sun shines, the snow sparkles with tiny points of brilliant light like minute coloured sequins. Biggish pieces of ice swing and sway over the surface of Arthur Harbour. The autumn equinox will take place in two days, Wednesday 20 March, at 3.16 local time. The sun sets earlier and earlier, and hours of darkness will soon begin overtaking hours of daylight. But the weather is at last quiet, the days calm.

This season's experimental satellite tagging of south polars has turned out difficult. Not many birds can meet the criteria of having a chick still on home territory and likely to survive. Transmitters were attached to two birds, but one had to be removed because it wasn't riding well; the other was got rid of by its wearer. Now Bill and Donna are trying to find new candidates at the far end of Humble.

A south polar chick lies quietly on its back in Donna's lap, bantam-sized, little claws coming out of its feet like sharp, shiny steel hooks. She plays with its legs. 'It's three to four weeks old. Its survival will depend on how crafty its parents are, how skilled.' The chick is getting bored, so she turns it over and strokes the little bill. 'Each chick is so different.' This chick has never been handled, there's no record of the parents so it won't be weighed, measured or banded. A pertinent mini-lesson in the reality of data sets. The parents fly at Bill, calling their husky low alarm call, a tremolo repeated to the same length, then drop just out of reach from his net and wheel up. Nothing works. But two skuas must be caught and fitted with black boxes, so we move on to a banded pair with two chicks. Bill lopes after one chick, Donna scoots for the other, and they are weighed, measured and banded. The older – passive but in good condition – is parked under a net weighted with stones. The younger – thinner, scrawnier, determined – is starving. It will most likely be killed by its parents or the older chick, and eaten. Bill: 'We'll find its band on the island next year.' Donna carries the scrawny younger chick up a hill, holding it out like an offering. It calls, high and constant, trying to attract its parents. Which don't come near. Bill and Donna confer and change tactics. They will go for brown skuas. So we head for Dream, where Pair 4 still has two chicks.

West of Palmer the ice cliffs rise in contorted magnificence. The ice looks as if it has been whipped into froth; it tumbles and writhes like the billowing tops of cumulonimbus caught in freeze frame. Huge, bursting with pent-up energy, ominous in its silence. There are no small puffs of ice fall, no slips of tumbling ice pebbles. Here the mega-big one is waiting – so big we are forbidden to go near. The waves created by an average ice fall across Arthur Harbour send surf clopping and swilling high around the rocks. The waves from this ice fall in waiting would swamp a boat.

But we do putter very, very quietly, talking in low voices, close enough to look through a flying buttress of an ice arch, cathedral thick, joining the main bulk of the ice cliff to a spur resting on rock. Here, in the years Bill has worked at Palmer, an ice-covered promontory has been gradually released and revealed to be a new island. Except, of course, it is old. It has been revealed before. I long to go ashore, to see with unsystematic eye if lichens have begun clinging, if mites are lurking.

The new island has a name, officially accepted in the *Gazetteer of Antarctic Place Names*: Fraser Island. After Bill. And the rocky islets rising off the point have an unofficial name: Patterdo Rocks, after Donna.

The day turns grey and noticeably colder, but Dream always is colder. The loosely piled, thin slices of grey rock clatter like china plates as we walk, and we kick them to keep the fur seals at bay. Furs are everywhere, mostly males, some getting to breeding point, with heavy shoulders and thick necks, tags revealing they've come from South Georgia. Their pink-red, meaty-smelling faeces scatter the ground. Small groups of moulting Adélies stand on the grubby slopes of last spring's snow banks. Adélie numbers are diminishing so rapidly on the five inner island study sites, Bill plans to work the two extremes of Biscoe and Dream, to maintain protocols. But access isn't straightforward. NSF have given permission for two fieldworkers and a tent on Dream next summer, for a week. It's ironic, comments Bill. Rules covering human waste disposal are strict in Antarctica, which adds to the logistics of the camp. And we spend the summer treading in animal waste.

The Frasers are stubborn, Donna remarks, as we track up another hill, finding another fur monopolising the top. But stubbornness still does not produce a brown skua in the net. Bill and Donna try one stratagem after another. Neither of brown Pair 4 will be caught, but at least the two chicks are weighed, measured and banded. And 134 fur seals, 50 crabeaters, 21 elephants, 2 Weddells and 1 leopard are counted as part of the marine mammals census done at every study site, every visit.

We ride back across smooth sea, distant mountains glowing a yellowish white along the peninsula horizon, all other colours subdued to soft lavenders and mauves, tender beauty failed by description. I sit on the rounded edge of our small boat just above the surface of the satiny sea. A fur seal rolls in the clear water as we go by, flipper up. My face is very cold. Just not too cold. And I think how soon I will be back in crowded, dirty London, so far from this place. How much I will miss it. How difficult to explain why to those who know me.

And Donna pays me a compliment. She says I'm not half-bad. She says I'm half a Birder.

Torgersen's brown skuas are the last resort, but a wide band of brash ice bars the way. Bill stands in the zodiac to assess the lumps and fragments and decides it's too thick. To my sorrow. And it's late. So we give up and return to Station by an alternative route via the ocean side of Torgersen, across to the glacier face and along a narrow channel next to the cliffs, past sizeable chunks of glacier ice and big, loose conglomerations of small chips. The first pancake ice is forming, thin grey circles lying on the sea's surface.

Next morning, 19 March, we find out about Larsen B. The news arrives via my journalist daughter Rachel, who copies me the wires as they come into her BBC London newsroom.

Directly to the east of us, across the mountainous spine of the peninsula, a massive ice shelf has been falling to pieces since the end of January. At first seen only from space, the evidence – icebergs calving off the outer edge of the shelf at an increasing rate – was scavenged from a NASA MODIS satellite at the National Snow and Ice Data Center in Boulder, Colorado. But now the news breaks around the world. Larsen B, a massive shelf of floating ice, is splintering into smithereens, shattering at staggering speed. A total of 3,320 square kilometres has gone, with the last 2,500 square kilometres collapsing into fragments and sliver-shaped small icebergs in a sudden catastrophic disintegration. Ted Scambos gives updated reports on the satellite images from Boulder. Argentinian scientists from a nearby base are the first to view the chaos of milling ice. Five hundred billion tonnes of ice shelf unhooked for ever, the profile of the peninsula terminally dug into, a deep scallop where the straight line of the ice edge used to run.

All the way along the eastern side of the Antarctic Peninsula, and further south from us at Palmer, on the western side, ice shelves reach out from the coast, each named, their boundaries defined and marked in on maps, their vast bulk extending by two or three times the apparent width of the peninsula's rock-bound ribcage. But in January 1995 the northernmost ice shelf on the east, and the top section of the Larsen ice shelf known as Larsen A, both suddenly fell apart. Nine-tenths of the two ice shelves collapsed in just a few days, refuting all existing maps. Instant obsolesence. The speed, and magnitude, forced a realisation that ice shelves could disappear at unimagined speed. The peninsula's fringing ice shelves had begun retreating in the 1970s, some suffering significant losses from 1986. Now the scale, and pace, were increasing. The link between the pronounced regional warming and the disintegrating ice shelves was solid, and scientists predicted further ice shelf break-up. Larsen B, the next segment of the Larsen ice shelf to the south, a 720-billion-tonne block of ice, would go, scientists said. Eventually. Some time in the future. Not now.

The news galvanises the world's headlines as evidence of global warming. It is the largest single event in the series of retreats by peninsula ice shelves over the last thirty years.

Earlier this morning Palmer's temperature was recorded as 0.0°C. Earlier, tiny points of snow dropped, then melted. Now there's rain, dripping and sloshing, miserable conditions. As I read the wires from London, the falls rumbling from our ice cliffs are so loud and insistent I check outside. All I see are the remains: clouds of ice smoke and ice crumble lying in the water.

Palmer is only 160 kilometres west of Larsen B, as the skua flies. We knew nothing about what was happening. Locked in our small, claustrophobic world, we could be a world away. But now, in our near neighbourhood, a massive ice shelf has collapsed at unimagined speed. It seems that this difficult, disastrous summer has claimed a major victim.

At dinner Bill is tired. He doesn't want to talk. All of us left-over summerers are pushing to finish our work. But he does tell me that I can say anything I want from this Palmer season. There are no constraints. I can use his results. Then he stands behind me, in the queue, silent, while we rinse our plates with the overhead power hose, and load the sanitiser.

Thursday the 21st. A Humble giant petrel fitted with a transmitter provides, by inference, evidence of the Larsen B break-up. On the return trip from South America it deviates from its expected flight path and travels down the east side of the Antarctic Peninsula over the newly released ocean.

Bill and Donna are still trying to catch a skua – any skua – and failing. The reasons are in the season, according to Bill. Chick loss has been so high, a distress call attracts a club hoping for food, and the parents are distracted. I'm tracking final interviews, checking my notes, getting all diaries and writing double-backed-up. Everything beyond my air ticket weight allowance is packed in strong, size-graded cardboard boxes for retrograding – sending back via the wondrous American system. One day, months ahead, having passed through all necessary procedures, the boxes will appear at my front door in London. Constructing boxes at Port Hueneme, California, is considered one of the least attractive ways into the US Antarctic Program. 'This job sucks' scrawled with feeling inside a lid. Packing induces uneasy feelings of finality, forcing the pace of departing. In only ten days the *Gould* will remove us.

On Saturday morning the rocks by Bio are strewn with the soft red of dead krill. An underwater cameraman filming krill for a BBC TV series showed me, last time I was at Palmer, how captured krill in the aquarium tank move defensively, with a rapid backward jerk of their bodies. Now I scoop handfuls of squirming little bodies from the seething intertidal pools, and the krill bump against my fingers, making heroic leaps for the water. Crabeater seals, heads sweeping back and forth just below the surface, are hoovering up the bonanza. Arthur Harbour is a krill harvest, kelp gulls floating gorged on the surface. Bill, matter-of-factly: 'It's typical, an intertidal beaching, it happens every March. There's a big bloom and the krill get caught as the tide goes out.' Then: 'Go into the field room. There you can see the story of the season.'

Three small piles from guano traps in Adélie colonies are drying on paper towel. Two have eggshell pieces stained pinky-brown. But the fragments from Litchfield are pale blue, clean, as if just laid, newly broken. Bill: 'Like little jewels.' Nests were abandoned so early on Litchfield there were no krill in the guano to stain the shells. Proof that the chicks never hatched. Random krill events aren't of interest to Bill. He wants data points, evidence. The eggshells give him affirmation of the disastrous year.

At last two brown skuas on Torgersen have submitted to transmitters. The new satellite work has been a bumpy, frustrating ride, taking an enormous amount of time. Bill: So many random things determined success or not. Everything was working against us. At least we learned that a much larger bundle of feathers is needed than we thought to make a nice tight package, not flopping. But skua feathers are very delicate and silky, more supple than giant petrel feathers. We didn't want to affect performance. In the end we have something to show for it: two tagged browns. And we have some positions where the two south polars tagged earlier were feeding, right over the Palmer Deep.

As for the Adélie satellite work – the weather affected every aspect: applying and removing tags, number of trips, survivability of the technology. The five functioning transmitters were switched between twelve birds over sixteen days, from 19 January to 4 February. Three tags were lost. One – on the male Adélie last heard of in the Joubin Islands, so symbolic of stress – Bill reckons just fell off. At least the tag I watched being fitted on the male with two youngish chicks on 26 January really delivered. Tag 14533 made four trips with its first owner, two with its second, before being recovered from a third bird.

The results? Twenty hours of data. Every single bird left Humble and headed for the Palmer Deep in a beeline. It's a common theme – birds seeking out foraging areas over fairly deep bathymetry. The Palmer Deep is in two sections, one fairly close, another beyond. They fed along the rim of the canyon in a semicircle. When they'd done feeding, they beelined on almost the same track back. The birds averaged 25.76 hours per trip, roughly a day to cover those distances. I could rush in and publish. Others might.' But, typical Bill· 'For me there's not much point. I'd rather have the long-term context.' The data provide further evidence of the significance to the local ecology of the Palmer Basin south-west of Palmer, with its winding valleys carved deep into the sea bed by the ice sheet.

All summer layers have shed from the roof of the great blue-shadowed ice cave in the ice cliffs, expanding the interior, lengthening the arch. We pause on the walkways, watching. No one wants to miss the finale. Further along the glacier face a slice detaches. For the beat of a second the hundred-foot-high segment hesitates, then collapses into itself. The sound cracks across the bay, a pistol shot out of sync, a silent movie with delayed soundtrack. Now, with immaculate timing, the end of the great ice cave comes as the last House Mouse for us summerers gets under way. The arch collapses, suddenly, in a smoke of crystals. A great roar, and it's over, cave obliterated. A river of pulverised ice disgorges down a new slope at intervals, the rumbles of adjustment sharp and intrusive.

Monday the 25th. Light snow, remarkably pervasive, covers most surfaces. The orange lichen is still bright on the side of boulders, but the colours are going, the islands turning to browny-grey and white. Brash lies over Arthur Harbour, blue-grey diminished. The second false fire alarm goes off in hours. Outside on the frozen road five people are digging up Antarctica, prising away small rocks contaminated by a burst hydraulic hose from one of the monster yellow vehicles. Twenty gallons covering 12 square metres, needing sorbent pads, a front-end loader, electric impact hammers and now shovels, resulting – so far – in five drums of pads, six and a half of soil and rocks, and a one-and-a-half-inch dent in the frozen ground. Jeff kneels, arms out, animated, telling stories.

The electronic version of the season's seabird data is finished, and archived in Excel spreadsheets in multiple places. The telemetry data is stored. Bill releases a brief summary for me. 'Here, baldly stated, are the facts of the season. For now, this data is private.'

The Palmer study sites have 7,000 breeding pairs of Adélie penguins.

In the summer season of 2001–02 there were 3,500 pairs. Fifty per cent of the Adélie penguins did not appear. But this is still a general figure.

This season Palmer had the heaviest snowfall and rain on record.

For the first time more one-chick nests were recorded than two-chick nests. 'This data tells us that the severe losses occurred at the egg stage. The environment eliminated the second egg. The snow was so severe it affected everybody. This year snow was an equaliser.'

One hundred and forty-eight Adélie fledglings were weighed on Humble – the lowest number ever, reflecting how few were produced. Seventy-five were weighed on Cormorant and Christine. 'These are the smallest chicks we have ever recorded. The average weight of a fledgling this year was 2,836 grams. The average weight of a fledgling Adélie penguin has been 3,035 grams. The 200 grams difference this season is very, very significant.'

Last year nearly 7,000 Adélie chicks were produced on the Palmer study sites. 'This year indicator counts have given an "off-the-cuff" census of 1,500.'

I need to understand the difference between the two measures of chick numbers: the standard measure, reproductive success data, and indicator counts, a 'Bill special'. Which leads to an explanation I've sorely needed. I'm reminded of a response I got when I asked the straight question: what are you trying to find out? Bill, tersely: 'It's in my articles.' Then he admitted it wasn't. 'The articles distil the research, in a particular way. The reasons behind it need explaining.'

The average long-term breeding success rate at Palmer's Adélie study sites is 1.34 chicks per pair. This year the reproductive success site data is .55 chicks per pair. But Bill thinks these data obtained by standard practice, highly labour-intensive, are not the best indicator of the ultimate consequences of what is happening. It is possible to have a constant number of repro success yet overall declining numbers. This year, indicator counts show a much worse breeding success rate than the repro success site data. 'I rely on them. I am here. I can see the population disintegrating.'

Indicator counts record when events occur. The result is a series of multiple bell curves accompanying the season, flowing through time. Bill: 'Indicator counts, done since 1989 in half the colonies, are the window through which we look at the timing of events in the breeding cycle of Adélies. It's what gives us a feel for how the season is going.' The database has columns of information, each column starting with 0 and ending with 0, beginning with the arrival of males at the colonies, then the arrival of females as the next column, then pair-forming, on with occupied nests, active nests, peak egg, a body count of chicks, crèching, the progression of departures as birds abandon the colonies and chick numbers decline. Each process, each procedure across the five-month season, peaks at a different time year to year. Everything is counted, every two days: 'No decisions, no judgement calls, just do them.' Bill: Their value is as a check, providing constant feedback into what is actually going on. If there are suspicious numbers, we can inter-rogate our other data. Our programme is always thinking of ways not to trust a simple number, ways to check on whether what we are doing is delivering accurate information. Never think that what you are doing is beyond question.

'Penguins are machines to some degree. They are under a timeline to have eggs on the ground by a certain point. They come equipped, loaded with supplies predetermined by the winter and the spring. Once they hit land, they have to perform within certain dates because they don't go off to sea to feed again. They have some flexibility – but how much? The window is narrow. The amount of time an Adélie can fast, for example, is hard-wired. Adults are more immune to environmental variability than chicks. But their ability to deal with it is preset to evolutionary parameters. Within them, the physiology of the Adélie can deal with x amount of variability. They breed to match peaks in key resources. Food availability within a season is finite, and declining. The hard wiring has a purpose. The ecological crunches this season have meant breeding late, struggling parents, lightweight chicks.'

Ecological crunches are delivered by weather. The impact of this ferocious summer has been played out in the small rocky colonies of Palmer's islands. It has been measured through the valiant efforts, the struggles, of Adélie penguins. Failed local lives. The evidence could not be clearer.

'This is an obvious place for a study of Adélie penguins', noted firmly in the biology notebook, Base N, 1957.

Yes. It has been.

'We understand that sea ice conditions influences the movement of birds and mammals.'

Yes. It does. Fundamentally.

On Tuesday 26 March the *Gould* arrives in calm seas and grey skies during breakfast. Shortcut is visited for the last time, and only eight south polar chicks are found worth banding on the whole island. Six more are still alive, small late-breeders, odd-looking. The season isn't a full reproductive failure for the south polar skuas. Not quite.

Donna clears the lab, making 'scolding for next year' notes: dirty weigh bags, an unblown dripping egg – all should have been dealt with. She chucks, cleans, packs, storing everything in the tent or the Grantees milvan, where it will freeze. She shows me the Haz Waste office in the next-door milvan, previously the Station shop, peaceful, warm, lined, with a window and a desk and a door opening to the sea. I never imagined Palmer had such a miraculous place for writing and thinking. Allocated, comments Donna, to someone who works set hours, mainly outside. The Field Room has to be cleaned. The tent has to be sorted. The aim is to keep others out. Bill: 'Sometimes we come back to find our space has been breached. There's no privacy on Station. The very bits you manage to pull together they violate.'

On Wednesday, squally, with soft fine snow covering our little world, collecting on the ground, rounding surfaces, the temperature is −3.7 °C. Palmer doesn't experience extremes of temperature, even in the depths of winter. The remnant snowdrift by the back walkway is now incorporated, so invisible. All vehicles have chains. Outside the window a shiny, post-moult Adélie climbs over rocks, footprints revealing its circuitous journeys. To the south, a line of small white icebergs marks the horizon, coming north.

The winterers are relaxed, about to get the Station to themselves.

At lunch we discuss names for people who keep coming back. OAE? No, because everyone claims to be an Old Antarctic Explorer. I suggest Antarctic Junkie? Maybe. Rocky: The first time you come for the experience, the second for the money, the third for the friends, the fourth because you can't fit in at home any more. He's arrived direct from McMurdo, with only two weeks away from Antarctica. And people laugh, in a comfortable way.

Bill walks up the glacier, looking at the new crevasses opened this summer. He reckons Gamage Point, the little promontory we live on, might be an island after all. Donna goes to Humble for the last weighing of her giant petrel fledglings. Someone has been found amongst the new winterers to take on the final stages of the work, recording the departure date of each fledgling. But one at least will never fly – four days ago the chick at 1.1 was found inexplicably dead in its nest. No weight loss, no appearance of starving. Donna: 'The father kept looking down then abruptly back up at me. It will be difficult to forget what I perceived as sheer confusion in his behaviour.'

The two final Adélie and giant petrel summer tasks are accomplished, both using satellite transmitters, both carrying on programmes begun a year ago. Donna attaches sacrificial tags 20470 and 20472 on two unbanded non-breeder females which happen to be on Humble attending play nests. Last year the two birds fitted with tags flew to the tip of South America, stayed, then returned to Palmer: 1,900 kilometres in six days. They worked up the west coast of Chile nearly as far as Puerto Mont, keeping close together, returning to Palmer between trips, until in mid-July, in the area of commercial tuna-fishing, the satellites stopped transmitting.

Bill fits tags on two post-moult fat healthy Adélies, a male and a female, from a group on Humble's big beach. The female gets the well-travelled black box 14533. Now it will start recording autumn journeys, part of the summer-winter continuity work. Bill: We watched the two birds return to their buddies preening on a snow bank, getting ready to settle for the night, heard them squawk, get a reply.

I do a last walk to the glacier, navigating snow-covered boulders and hollows by treading in the tracks of others. Then, snowclouds heavy, gaze my last gaze across our low lines of rocky land and sea to the distant Antarctic mainland. T5's owner, studying the constantly shifting views from the porch outside his office, tells me that on days when light conditions elevate the horizon – Fata Morgana, refraction of light in heated air layers – he has noted mountains 190–240 kilometres to the south.

Explorers of Antarctic seas on occasion insisted that they saw land where there was none. 'Appearance of land' marked down on charts, later denied. But I have seen, from the bridge of the *Gould*, 130 kilometres from the coast, a snow-covered mountain suddenly appear on the horizon richly golden, like a roundel in a stained-glass window, then disappear.

The icecaps covering the islands along the coast of the continent glow a soft lilac. Then, rising behind, two enormous mountains materialise, immensely rugged peaks gleaming, solid, real, in brilliant sunshine. Nothing else. Just black rock, white ice, two giant bony structures.

Then they are gone. I need no other farewell.

30

Voyage home

Friday 29 March to Thursday 4 April

We leave Palmer early on Friday morning, 29 March. No one is inclined
to 'jump the bumper' – board after the gangplank is up. Our departure
is subdued. Yesterday went in a jumble of rushing: room cleaned, bed
linen washed, down comforter washed, luggage packed and into the
Float Coat room by 12 noon. Everything off shore and on board the
Gould by 3 p.m. Find my berth, stow possessions against the Drake,
stack weather gear, boots, thick socks, gloves, neck-warmer, hat, at the
ready; to sleep my last Palmer night in a narrow *Gould* bunk, fold my
height once more into the brief gap between lower and upper berths. My
journal has just three terse notes: −4.5°C. Clear and sunny. Extremely
tired and hassled.

For those of us up on deck there's a copybook beauty ride through
the inner passages. A rumour lingers that we might, as a kind of com-
pensation for a hard season, visit our neighbour Port Lockroy, old Base
A established in 1944, restored by fund-raising to its 1962 state and
now the most frequented tourist site in Antarctica. But it dissipates in
the reality of the *Gould*'s tight schedules.

The sun sets in a red sky. Red sky at night, sailor's delight: 'Any
correlation?' I ask Jay, padding around the ship as first officer. 'None.
The same with whistling being bad luck,' he says. 'I do it all the time.
Silly superstition,' and he slopes off, whistling.

Once in the Drake, with lumpy seas and heavy grey skies, almost
everyone feels low, if not actually sick. The trip is unpleasant in a dif-
ficult-to-define way except that we are all tired, flat, just wanting it
to be over. But, as ever, Bill and Donna can't stop. The first of this

winter's GLOBEC cruises will depart from Punta two days after we arrive, taking Brett and Chris to carry on the summer–winter continuity work in Marguerite Bay. This time they'll camp for twelve days on Avian Island, south of Rothera, diet sampling Adélies and attaching satellite transmitters. Logistics need to be sorted: the computer screen will self-destruct if the tent isn't heated. Bill: 'I feel sometimes I have one continuous field season. But winter time makes quantum leaps in what you know.'

As we head north, satellites pick up signals from six small birds pinpointed in the great Antarctic emptiness: two brown skuas, two giant petrels, two Adélie penguins. Each, invested with a precious transmitter, venturing out into sea and sky. Except that one brown skua is still firmly on Torgersen. The other is due east from our current position, in the Falkland Islands. One giant petrel has flown the same route as last year's birds to the same first landfall, an island south of Cape Horn; the second is following in the same track. I think of Donna, eager, pressing: 'I need bigger numbers to be credible. I'll only have four transmitter results for non-breeders at the end of this season. If I show them at a conference, I have a big 4 on my chest. In my perfect world I'd put out twenty transmitters a year. But twenty transmitters cost $60,000.'

The female Adélie bearing tag 14533 is currently 35 miles south of Palmer in the Argentine Islands, old Base F territory. Bill tells me she's leaving in the morning, feeding in the Palmer Deep around midday and coming back in the afternoon. 'My interest is not only where the Adélies are feeding, but what they are doing.' Does he worry about tags constraining the birds? 'No. They currently have much less pressure to meet a schedule. The distances they have to cover from their feeding areas to their night-time haul-outs where they roost is 10–15 miles. So there's no big problem.' Tag 14533, programmed to deliver information once every four days to conserve batteries, will with luck have enough transmissions for four or five months, till the end of winter.

The *Gould* grinds on. No one much bothers with the lounge; the decks are empty, the galley largely unattended.

I've been thinking about the influence of the sea on the work at

Palmer. Bill and his team are separated from their places of study by a boat ride. They must be daily, almost hourly, aware of the sea's condition, assessing, experiencing. I suspect that this intimate, unavoidable relationship has contributed to Bill's breaking out of the box – to his moving beyond the nesting colonies and working on Palmer's Adélies in the wider context.

Bill says yes. This is a truth. Plus the sea prevents actions, forbids definite plans. It's not possible to count on maintaining a schedule. 'Perhaps that's why we have developed ways of double-checking records by indicator counts. The sea has driven our database, put unpredictability into our way of doing things. We've had to develop a way of compensating. The sea, subconsciously, gives us a feel for how things are.'

What about the year ahead?

Bill: The big mystery is, what is going to happen to the sea ice this coming winter. When, where, how heavy? I predict ice as far north as the Drake. What happens this coming winter in terms of ice and snow will be the definitive environmental event that will determine very quickly what the status of these Adélie populations will be in five to seven years' time.

How many penguins will return? Who lived and who died? If anything holds me in suspense, it's this. There are scientific reasons. And a sense of tragedy. This is as close to an emotional issue as I can bring to bear on this work. It's quite possible that 50 per cent of the birds we were working with last year are dead. The thought that those might be gone ... But you could say the adults called it off, took a sabbatical, and they now have more energy for the winter. They have a long lifespan, maybe twenty years. The ideal model is that every bird is replacing itself. Except, climate change is changing the rules.

Look at how the Adélie populations have decreased here. A drop, then it flattens, then drops, then flat. This season's fledgling weights might represent the beginning of the end. I'm leaving here truly apprehensive about these birds' long-term future.

'How do ecosystems respond to climate change? Gentoos and chinstraps are increasing here at Palmer. Open water, less ice, means different species invade. But it's not inconceivable that with no sea ice the krill would vanish. Gentoos eat fish and krill, so are susceptible to the same changes in the ecology. They can handle substantial variabilities. But how much, and how fast a rate of change? They will probably overtake Adélies in the next twenty years.'

And he repeats: What we need to know about is happening at sea. Where Adélies are wintering, where feeding. The survival of the fledglings. This is truly the key.

Me: Will you do this?

Bill: 'No. I won't. I'm looking for my successor. But we have identified the next adventure in this research.

'Adélies are the window by which I look at the system. They are a means to an end. I love Adélies.' A pause. 'I feel for them. I feel compassion for these animals.'

Bill has only recently told me that he still thinks in Spanish, translating in his head. He does all his maths in Spanish. 'I find making speeches difficult. Talking one-to-one is easier.' Bill's inclination is to be private, but our conversations, chiselled out of his tight schedules, extracting time when he was able to give it, have been central.

The last day, the day before arriving, we sit in one of the big empty labs, swells sloshing half-up the portholes, engines thumping, a couple of postgrads, research completed, perched cross-legged on lurching lab benches, trying to read.

Birds and animals, Bill explains, can bear more variety in summer than in winter. Winter is the crunch time. If the winter habitat goes, there are very immediate and severe effects on populations. Now the winter habitat on which the penguins at Palmer clearly depend is disappearing. It is indispensable to their survival. He summarises: the mechanism for climate warming affecting Palmer's penguins are fairly straightforward. There are only two. On the larger scale, vanishing sea ice. And, on the smaller scale, landscape. The effects of snow and rain.

A pause: 'It is possible that these two are dead wrong. But these are the two that seem to me to make most sense right now.'

'Adélie penguins are among the region's top predators. They are intimately linked to sea ice, and to krill, which pre-qualifies them to be pretty good indicators of climate change. The Adélie work has been the start of understanding a cascade effect, and it has developed as part of that cascade. Evidence from the Palmer Adélies integrates into larger spatial and temporal scales.

Major ecological events are usually measured at small scales first, where the signature is usually picked up. It has to be visible somewhere. Look at what happened this season. The magnitude of the changes. The heaviest snowfalls. The biggest rainfalls on record. Fifty per cent of Palmer's Adélies didn't turn up. The failures among those that did. The grand finale, the collapse of Larsen B. This region could be a very, very sensitive indicator.'

The *Gould*, half empty, summer over, labs untenanted, is curiously bleak. Bill: 'It's been my longest period of time here, since I wintered aged twenty-five. So much time in the field.

'It was a bad, mad year. The most astonishing season I've ever had.'

Landed in Punta Arenas, brief bonds dissolve. Down we went, and back we've come, to go our own ways.

A pack of Punta's dogs trots across the main street. The usual random mix of big, smaller, black, brown, patched. There's a sudden burst of barking and tooting. The traffic stops. A dog is in the roadway, alert but immobile, front legs stuck forward like a stone dog on a medieval tomb. The lead dog puts a foreleg either side of the dog's hindquarters, hunches down, and drags it to the side, where the rest of the pack stand quietly.

Antarctica, people claim, is a place where we watch out for each other. We could learn from Punta's dogs. Once included in the pack, no hidden exclusion clauses.

I go into a supermarket to buy fruit and stare, shocked, at a fly, walking across the stacked apples. And, at Santiago, in the concourse of the international airport, a shock big enough to make me jump: a man wearing a suit.

❄

Ten days after reaching London I fly down the world again. On a wild afternoon on the western Australian coast, autumn rain pelting in, a gale blowing out of the south-west, I see the browny-grey plumage of a large bird tossed like ash in smoke above the heavy surf, glimpse it exhausted through the sheets of rain, start running down the sand hills – then lose all sight of it.

I couldn't be certain. But I sense it was a giant petrel.

3 1

What was happening?

Shakespeare's *Coriolanus*, a play of power and civil conflict, starts with rebellion. The hungry people of Rome are mutinous. The patrician, Menenius, calls their leader, armed with a club, 'the great toe of this assembly.' The great toe, says Menenius, is the lowest, basest, poorest part of the body – yet it goes ahead, first.

A climate rebellion was happening in the world's great toe, the Antarctic Peninsula. Poor, base, low, by the world's measuring, insufficiently acknowledged in climate change accounting. Yet it was warming fast: the Antarctic Peninsula was going ahead, first.

How did the changes in this extremity connect with the body – to the rest of Antarctica, and to the rest of the world? Scientists would not commit. The causes for the rapid warming on the Antarctic Peninsula were not known. The profound warming in this remote place could not be linked to anywhere else.

John King, head of the Antarctic Climate Processes programme at the British Antarctic Survey, gave me advice. People will want you to connect what is happening on the Antarctic Peninsula to global warming, to say it is evidence, he said. You can't really say this. The west coast of the peninsula is probably the most rapidly warming region in the globe. But the link between the temperature figures, and the events on the Antarctic Peninsula and in the wider world, can't be made. 'Correlation does not necessarily indicate causation.' The toe did not connect.

In September 2004 scientists from around the world gathered at the Scott Polar Research Institute in Cambridge to present the latest research on Antarctic Peninsula climate change. The focus was on Bill's bad, mad year; in particular September 2001 to March 2002, the ferocious summer.

And this is what, it turned out, had happened.

A totally anomalous atmospheric circulation pattern began late in September 2001. A strong low-pressure system backed by a blocking high settled like a collar around the top of the Antarctic Peninsula and did not shift for more than five consecutive months. A strong, north/north-westerly flow of mid-latitude air, mild and moist, persisted over the northern Antarctic Peninsula from early spring, through summer, to the start of autumn. Temperatures rose. Snowfalls increased, and rain – wetting, temperate rain – fell in unimagined quantities. The rise in surface-air temperature was unprecedented. It was the warmest summer on record on the Antarctic Peninsula. The impacts were profound, and complex.

First and most visible, was the sudden, disastrous disintegration of Larsen B in February and March 2002.

With surface air temperatures increasing over the peninsula, the melt season had been gradually lengthening. Ice shelves had been retreating for several decades, warmer summers trimming a few kilometres from their fronts. Ponds of melt water had been appearing on the surface of the northernmost ice shelves. But now, in the exceptional weather pattern of 2001–02, Larsen B experienced intense and prolonged surface-melting. The water, collecting in ponds, ran down rifts and existing crevasses into the body of the ice shelf. Crucially the melt-water ponds were constantly replenishing, staying brim-full, maintaining the flow of water into the cracks. With constant pressure on their sides and tips, the cracks extended and deepened.

It was the weight of water that fractured the mighty mass of Larsen B. Penetrated with holes like a Swiss cheese, the ice fell apart, suddenly, at speed. Not calving in the usual blocks, but splintering; icebergs in slivers. As the front of the ice shelf began breaking away, as its body cracked apart, ice milled in flotillas of clotted bergs and brilliant blue brash.

By the end of the third week in March the northern two-thirds of Larsen B had gone. Ocean ran where only eight weeks earlier 220 metres' thickness of glittering white ice had defined the outline of the Antarctic Peninsula.

Summer temperatures in the peninsula had been just a few degrees below what researchers believed was the threshold for surface ponding, and ice cracking events. Now, with the peninsula's profound climate warming, a threshold was being crossed.

Second, was the impact of the anomalous weather on sea ice – complex and profound, according to the scientists. Along the peninsula's eastern side the strong persistent winds swept away the heavy sea ice that normally protected the seaward facing edge of Larsen B. Sea ice buffer removed, Larsen B was exposed to open-ocean conditions, hastening its break up. On the western side of the peninsula the strong persistent winds forced the sea ice against the coast. Pieces stacked on top of each other, compacting, deforming, causing immense thickening, accretions of ice clumping deep in the water. The repeated storms and warm moisture-laden winds dumped a significant cover of snow over the sea ice, but the high temperatures caused unseasonable ice and snow melt, freshening the upper ocean.

Sea ice extent was decreasing in the Western Antarctic Peninsula region, the annual ice forming later, retreating earlier, shortening the overall sea ice season. But in 2001–02 the region's sea ice cover rapidly advanced then retreated early and fast, resulting in an exceptionally low seasonal extent. However the unusually heavy sea ice conditions lasted throughout the summer. Scientists realised that the established descriptor of sea ice – its extent – was inadequate.

A third impact of the weather was the unexpected effect on glaciers. Researchers tracking the breakup of Larsen B now realised that ice shelves act as door stops, plugs, to the glaciers that feed them from the land behind. The break up of an ice shelf can impact on those glaciers – balance destroyed, no longer constrained, they can respond rapidly and dramatically to the departure of the fronting ice. Larsen B as an ice shelf was of course already floating and its collapse did not add to sea level. But ice from glaciers – ice from off the land – does add to sea level. Glaciers behind the departed Larsen B were shifting their loads up to five times faster from the interior of the peninsula to the sea. The image scientists used was simple: the cork in a bottle. Remove it and the contents flow out. With Larsen B gone, ice from off the peninsula was falling at increasing rate and speed into the ocean.

Scientists were eager to discover if Larsen B had broken up any time in the past. What could they learn about the long-term climate history of the peninsula? To the north, the sprawling Larsen A ice shelf that collapsed in 1995 turned out to have a history of growth, fracture and renewal. Researchers investigating the newly revealed seabed beneath Larsen B found no evidence of any previous breakup. Larsen B's collapse was a one-off, and final: an ice shelf had been here for at least the last 12,000 years. And the Jumbo Piston Corer we ducked under as we climbed on board the *Palmer* that stormy February afternoon in 2002 delivered more climate history data. Sediment cores showed that the outer edge of the vast ice sheet once covering the Antarctic Peninsula and its fringing islands had reached out 200 kilometres from land, almost to where the continental shelf plunges down to the ocean depths. The ice sheet had started pulling back, retreating, around 16,190 years ago.

The fourth impact of the anomalous weather pattern was on the Antarctic marine ecosystem, intimately coupled as it is to the annual advance, retreat and behaviour of sea ice cover. Impacts were seen at all levels. An unexpected major phytoplankton bloom occurred; but krill, needing sea ice as a refuge, and a feeding ground under the ice, 'did not', in cautious scientist-speak, 'appear to benefit'. The effect on the local Adélie penguins was as I already knew: they suffered – in the regional context of sea ice, and locally, via heavy snow. Significant changes in penguin populations occurred.

Bill was at the Cambridge workshop but he did not present a paper. At Palmer we were reluctant recipients of the extreme, anomalous weather conditions. We could not anticipate the time-scales. We lived among the results as they unfolded through that difficult, confusing summer of 2001–02. Inherent caution is central to Bill's approach.

Conclusions come from a matrix of trials, alternatives, observing, cogitating and waiting for results. 'Now you know', he said 'why I do not rush to publish.'

Other pieces of Antarctic Peninsula warming slotted into place. Seasonal snow cover was shown to be shrinking. None of the seven fringing ice shelves that had already collapsed had re-formed. Researchers using historic aerial photographs combined with current satellite images revealed that coastal glaciers, starting with glaciers in the north, and moving south, were in overall retreat. Glaciers are generally reckoned to respond slowly to climate change, and shifts in their status are important indicators.

The signals were dramatically clear. Strong threads, observed, measured, were coming together. Climate-related processes could result in near instantaneous ice shelf losses, scientists concluded. The sudden disintegration of Larsen B ice shelf was climate-induced. In addition, the peninsula's rapid warming was causing the decline of the Adélie penguins in the Palmer area. Time-scales of sufficient length were in place for both events: Larsen B had been an ice shelf for at least 12,000 years; Adélies had nested at Palmer for the last seven centuries. Neither event could be considered part of a natural variability in the region. Both were evidence in an evidence-sparse, complex, fast-moving world of climate change.

But scientists still did not know what caused the rapid warming. The mechanisms, the means by which change is effected, were still uncertain. Climate models can be tested by backcasting, by running them to see if they predict changes that are known to have already occurred. Increasing amounts of new data were being added to the models. But none could reproduce all the observed facts of warming on the Antarctic Peninsula.

The ferocious summer had stoked the rebellion to a new intensity. But the big toe still could not be connected to the body.

I needed the final figures for the 2001–02 Adélie season at Palmer, the results from the five inner-island study sites. The figures Bill had given me as we left Palmer in March 2002 were unofficial.

The 2004–05 Antarctic season was about to begin, with the annual migration of scientists south: northern autumn to Antarctic spring. As always, Bill was immersed in the intense activity of the seabird ecology work at Palmer. But at the end of March 2005, with summer over, the season's Adélie chicks fledged, databases in order, Bill sent me the figures, using proofed data as in his archives.

In October 2000, at the start of the 2000–01 season, a total of 7,161 pairs of Adélies arrived to breed at the five Palmer study sites. A year later, in October 2001, anomalous weather underway, 4,288 pairs of Adélies returned to the study sites to breed. A startling drop, representing a 40 per cent decrease in the between-season breeding population: the largest since records began.

The following season, 2002–03, was a year of heavy winter sea ice, as Bill had predicted, extending all the way to the Drake, with amongst the lowest snowfall. Five thousand, six hundred and thirty-five pairs of Adélies appeared at the study sites. So some of the missing Adélies had returned. But Bill calculated that more than 1,500 pairs of Adélies from the Palmer sites must have perished during the winter of 2001. Given the small size of the population from which these were drawn, the 1,500-plus pairs represent a substantial number.

There had been only one previous large annual drop in the Adélie data set: a 28 per cent decrease during the 1987–88 season, coinciding with the timing of an ageing krill population. To Bill the figures for the 2001–02 Adélies were evidence of another such event. The last great krill cohorts, previous to 2001–02, were produced consecutively in 1995–96 and 1996–97. The survivors would have been five to six years old by the season of 2001–02, their abundance greatly diminished.

Bill: 'What seems clear is that a lack of krill not only increases the over-winter mortality of Adélies, but also forces some birds into making the "decision" not to reproduce, probably due to poor body condition.'

Assessing the data at the Cambridge workshop the previous September: 'In my mind at least, 01–02 is starting to look like Adélie hell. No ice, too much snow, no krill.'

Speculating about population trends was now less clear. In 2003–04, 5,350 pairs of Adélies appeared at the Palmer breeding sites. But the confounding event of the following winter of 2004 was that it was one of the heaviest ice years – if not the heaviest – since 1980. Too much ice can be as detrimental to Adélies as too little. At Palmer, despite an apparently good recruitment of krill, the Adélie numbers dropped, with 4,798 pairs arriving for the 2004–05 season. Bill made the assumption that it was the heavy ice that depressed the population.

The reproductive success rate for the 2001–02 season was 0.78, the lowest since data collection started. Only 1,729 chicks were produced, compared with 6,531 the previous season. But Bill had some optimism about the ferocious summer chicks. Although they were lightweight, they turned out to have fledged into a winter of strong ice and increasing krill abundance. In February 2002 Bill had given me a cut-off point below which fledging Adélie chicks were fatally compromised, of 3,200 grams. But these numbers are adjusted a little every time new data are added. The latest analysis gave the average weight of chicks that did not survive as 3,035 grams. 'When conditions are "right" – lots of ice and lots of krill – even marginal chicks may survive', Bill wrote to me, 'and this may be the key role played by sea ice in Adélie survival – sea ice produces krill – and it also gives Adélies the platform from which to access krill during the critical first winter of their life. It is a difficult issue to disentangle. However, it makes sense to me instinctually and within the bigger picture of the ecology/evolution of Adélies.'

Wrestling as always with the shifts between local and regional, the scales of time and place implicit in studying penguins, Bill was starting to look at the short time-scale of ecology in relation to the long time-scales of geology and glaciology. Adélie penguins breed in vast colonies on the coasts of the continent. But along the entire convoluted length of the Western Antarctic Peninsula, coastlines glacially sculpted, cut with islands, bays, fjords, deeps, Adélies breed in only five locations. Each is associated with geological features gouged out of the sea-bed by the most recent ice sheet, deep undersea canyons cutting across the continental shelf, with upwelling Circumpolar Deep Water encouraging the growth of phytoplankton grazed by the krill and fish on which Adélies depend.

Bill: 'Adélies – swimmers not fliers, requiring daylight for foraging – need adequate food supplies, in accessible places, to do the two things they need to do: survive winter, and breed successfully.'

In summer Adélies breed in areas tied to predictable prey availability. In winter they go to areas defined by the same parameters. Palmer's penguins were being hit by the loss of their haul-out platform, sea ice. They were being hit by increasing snow at nesting sites, by drowned eggs and predation. But accessing food, where they need it, when they need it, is key. A warming peninsula was forcing a mismatch between the resources Adélies need and their ability to access them. It was having an impact on the abundance and availability of prey.

Bill's Adélies, struggling, failing, persisting, succumbing, were a small poignant example of a potentially vast reality. In one sense, they had become surrogate humans. Through them, the impact of a changing climate on established communities was palpably visible, a kind of parable in real time. This is how it is, if you have been residing in attractive island real estate, and climate change comes knocking on your door.

At start, and finish, Palmer's penguins as indicator species.

The lack of mechanisms permeated the science of climate change in the Antarctic Peninsular. Scientists grappled with climatic oscillations or see-saws, cyclical changes over years, or decades, or centuries, with local reach, or regional, or hemispheric. Sentences were embedded in acronyms and insider shorthand. Conclusions were shored up with caution: 'possibly', 'may have', 'might'. Climate change is a fraught subject, tangled in huge issues, scientific and political.

Current human actions contributing to the warming of the planet were not in question. But without agreed mechanisms the pressure to understand current climate processes, to try and predict the future, could not be met. Some scientists still worried that the abrupt climate changes happening in the Antarctic Peninsula could potentially be within the natural range of past climate variability. So little was known about the recent climate history of the region. Geologically there had been cooling interrupted by warmer periods. The world's high glaciers in the tropics and subtropics were melting at an incredibly fast rate, mainly from increasing temperatures. Data from ice cores revealed that the warming at these high altitudes was unprecedented during at least the last two thousand years. But equivalent data did not yet exist for the Antarctic Peninsula, where few cores had been taken – although more were planned. Two drilled in 1990 gave a date for the onset of warming in the 1930s; but no information was available further back than 1510, the depth reached in the ice by the deepest 235-metre core.

Behind everything was that intellectual challenge, the unsolved riddle, the preoccupying puzzle: what drives the temperature? Why does the climate change? It has changed. It does change. It will continue to do so. Why?

George Denton, classical geologist, talking about glacier retreats in New Zealand in January 2006: 'We are having a big warm event now. The glaciers show a big retreat, a small advance followed by a big retreat, another small advance. An advance is short-term. Retreats are long-term ... I'm not interested in global warming. I'm interested in climate change. Climate change is exactly that. Change. What I want to know about is what triggers ice ages. That is the great unanswered question. We are in an interglacial now. How long?'

For scientists the dream, the driver, is of a solution being embedded within their own particular areas of study. For the majority of researchers living and working in the northern hemisphere, it's a raft of potential triggers north of the equator. For the few who go south, deep south, it is the role of that generally ignored, insufficiently studied, enigma, Antarctica. How does Antarctica operate as an engine of climate? What is its significance in current climate change?

But events were moving fast in the world's great toe. Drastic changes were occurring. Rising temperatures with prolonged melting conditions were profoundly affecting the peninsula's ice sheet. Permanent snow cover was reducing, snow and ice banks disappearing, areas of bare rock extending. Glaciers retreated at an accelerating rate. Changes to the ecosystem were dramatic. The polar system was disintegrating. Sub-Antarctic ecology replaced the Antarctic at increasing speed. Ice-intolerant animals replaced the ice-dependent. Plants were spreading south.

Researchers monitoring the area vacated by Larsen B reported that glaciers previously restrained by the ice shelf were flowing unrestricted into the ocean at an ever-increasing speed. Their fronts, ice shelf gone, were being additionally stressed by warming ocean water and storm action. To the south the next ice shelf in line, Larsen C, was noticeably thinning.

Adjustments to the detail of the peninsula's intricate spaces went, by definition, almost all unmarked; but along our local Palmer coast-line there was another new island. On 10 January 2004 at 19.15 local time the last of the ice arches linking Norsel Point to Anvers Island collapsed, sending an explosion of ice and water surging into Arthur Harbour. In 1955 six young men, newly arrived on Anvers Island, had hauled their sledge from the backdoor of the Base N hut up the long ice ramp on to the piedmont, a 2,000-foot climb. Now, behind the foundations of the hut, there was rock, glacial gravel and bastions of remnant ice melting, according to Palmer's summer manager, like a dropped ice cream. Perhaps 200 feet below the slope of the 1950s ice ramp, today's rocky surface ends abruptly in a deep channel running with ocean tides. And behind our Station buildings, the glacier was now revealing a new crevassed depression, a potential break-line where Gamage Point could perhaps one day disengage from Anvers, stranding Palmer Station on yet another small island.

Along the western side of the peninsula significant changes had occurred to the sea ice. Annual mean extent had reduced by 40 per cent over a twenty-six-year period scientists reported, at an Antarctic Peninsula workshop in Boulder, Colorado, May 2006. Ocean surface temperatures had risen by more than 1°C between 1955 and 1998, and salinity had increased. Sea ice continued to advance later in winter and retreated earlier in spring/summer. Decreased concentrations resulted in warmer, saltier sea water, further reducing sea ice, sustaining and enhancing the warming. A positive feedback.

Climate change varied with location. The north of the Antarctic Peninsula was different from further south, the western side from the east. On the west the dramatic warming had occurred in midwinter: according to the latest calculations a 6.3 °C increase in July temperatures since 1951, the most rapid winter warming measured anywhere on the planet. Overall, since 1955, the increase in annual surface air temperature here in the Western Antarctic Peninsula was almost 3 °C, approximately ten times the mean rate of global warming. Experts estimating the temperature increase the planet could experience described a warming of 2 °C above pre-industrial levels as dangerous. Palmer, and the Western Antarctic Peninsula, had experienced a rise of 3 °C in just fifty years.

But the mechanisms for warming on the Western Antarctic Peninsula were not yet understood. Despite decades of effort, the ways atmosphere, ocean and sea ice interacted in this complex environment were still unclear.

The eastern side of the peninsula had always been markedly colder and drier than the west. Antarctica's two flowering plants never managed to get established anywhere on the east, not in the last 10,000 years. But now the northern part of the eastern side was experiencing rapid warming, and the warming was happening in summer. Wind patterns had altered. There was a significant increase in the strength of the west winds blowing across the peninsula, particularly in summer, the months when Larsen B's disintegration had taken place. Since the mid-1960s near-surface temperatures had increased by more than 2 °C.

And at last, at least for this part of the Antarctic Peninsula, a mechanism for warming had been established. Robust models linked the altered wind patterns to stratospheric cooling resulting from ozone depletion over Antarctica in autumn and increased greenhouse gases in the atmosphere. Observed changes fitted with what the model predicted. The warm summers on the east coast were human-induced.

Since the ferocious summer, changes on the Antarctic Peninsula were occurring faster than scientists had ever anticipated. They were of greater scale, speed and magnitude than had ever been considered possible. Warming so fast, so intense, so widespread, was shocking. The speed was truly significant, the impacts profound. A consensus emerged. The rate of atmospheric warming was now agreed to be unprecedented in the recent geological record. What was happening in the Western Antarctic Peninsula marine system was a premier example of rapid climate change, as experienced globally, declared the key Palmer LTER scientists including Bill, in a paper published in January 2007.

John King, when I interviewed him at the end of 2006: 'The warming in the Antarctic Peninsula is part of global warming. You can connect the Antarctic Peninsula to Antarctica – and to the world.' John is renowned for being a careful, cautious scientist. But now he was convinced.

The long uncertainty had reached a natural conclusion. Poor, base, low by the world's measuring – but the great toe was going ahead, first. Climate rebellion in the Antarctic Peninsula was part of the world's rebellion. The toe's warming was the body's.

The latest maps of the Antarctic Peninsula show a large area of ice sheet: the Larsen. The coast above it bears no evidence of the recently departed, of Prince Gustav Ice Shelf, Larsen A or Larsen B. They were, but they no longer are. A new map of Palmer Station and its islands, commissioned for this book, shows Norsel Point detached, no longer part of Anvers Island.

Science is driven by what currently is. Seamless shifts in the expression of certainty. Adjustments of conviction. The latest data preempts previous data. Historians work to discover what was, to examine the seams. To uncover the ideas and arguments, the struggles, the uncertainties and the evidence for what was, until superseded, current.

But historians never conclude. They just pause. Perhaps history and science are not that far apart.

32

Local becomes global

The Antarctic Peninsula is unstitching. Ice slides off, crumbling, tumbling. Glaciers shrink and thin, discharge rates accelerating. Ice shelves are at risk, melt water penetrating consistently, deeply, into ancient structures. Along the western coast the sea's annual coating of ice performs fitfully.

The stitches were held together by cold. Warmth denies their performance, removes their ability to function. The Antarctic Peninsula was cold. Now it is warming. Temperature, and the seasons, are everything.

The ecology slides down the peninsula like the skin off a snake. Underneath is a new skin. But there the analogy ends. Adélie penguins are being sloughed off. The peninsula's new skin is attracting different penguins: chinstraps and gentoos. Polar Weddell seals are being replaced by sub-polar fur seals and elephant seals. The new skin has increasing numbers of plants. It includes non-indigenous microbes and fungi. It has introduced uncertainty to the inhabitants of the surrounding seas: cold-water marine organisms are peculiarly vulnerable to temperature change. The two significant prey species are affected. The regional abundance of krill is in decline, and larval Antarctic silverfish are no longer found in northern waters.

As predicted by the earliest climate models, dramatic changes are occurring in the far north of the planet and the far south. There, and on the upper slopes of the planet's high mountains, species are most at risk. Cold terrestrial regions have fewer species. There are not many places left to go.

It takes surprisingly little warming, or cooling, to make a difference to climate. But in parts of the Antarctic Peninsula temperatures hover around a critical divide: the melting-point of ice. The crucial shift is from freezing to liquid, liquid to freezing: that extraordinary dual state of water, tipping either edge of a point, flipping function and status within the smallest range. The peninsula is now a permanent performer in the theatre of ice to water. It has begun to decouple from the ice age.

Ice parks water. Ice ages take the oceans and redeploy some of their contents as ice, releasing a greater surface area of dry land. Not-ice ages take back the ice, as water, making do with less land. Changes in the volume of ice means changes in the volume of the oceans. Changes in the volume of ice occur because of changes in global temperature. If it gets too warm, ice melts. The physics is very simple. There's no argument. The quantity of ice increases with cooling, decreases with warming.

The planet's most recent ice age ended with mid-latitude warming. The ice at mid-latitudes, with the exception of glaciers on high mountains, melted. Flooding was widespread and catastrophic. Sea-levels rose by almost 100 metres between 14,000 and 6,000 years ago: on average a metre a century.

The planet's high-latitude ice shrank, but enough remained. Thirty-three million cubic kilometres of fresh water is still stacked in massive ice sheets. The northern polar region's stores are in that remnant of the last ice age, Greenland. But it is in the south, in Antarctica, that most of Earth's ice is banked. The Antarctic Peninsula holds comparatively little of the account. If all its mountain glaciers melted, global sea-level would go up by only one third of a metre. The vast bulk lies over the polar plateau in the East Antarctic Ice Sheet. Below the peninsula's roots, the West Antarctic Ice Sheet has the rest. The great polar ice sheets contain the equivalent of seventy metres of sea level rise, if added to the planet's oceans.

The melting of polar ice is the largest potential contributor by far to diminishing land surfaces; to where, exactly, the incoming tide might reach.

The expectation has been that current global warming would cause increased snow precipitation over the polar ice sheets. They would gain mass. Which is where I came in, learning about the Lambert Glacier Traverse in 1994. The results from the traverse set the baseline for estimates of Antarctica's contribution to sea-level rise. In 2001 the UN Intergovernmental Panel on Climate Change (IPCC) Third Assessment Report gave the contribution from Antarctica over the previous century as minimal. As for the next hundred years, scientists predicted that Antarctic ice sheets would probably gain mass: their contribution to sea-level rise would remain small.

But despite increased snow precipitation, despite expectations, the great polar ice sheets are losing mass. In Greenland the ice sheet is experiencing rapid near-coastal thinning. In the Arctic ocean, summer sea ice thins and shrinks, faster than computer models have calculated. And Antarctica is leaking. Earth's high latitude ice is not inviolate. That coating of intense white, those two patches of reflecting crystal cold, are not a given.

Substantial improvements in methods for monitoring ice sheets have occurred. The bamboo markers and towers of fuel drums, the physical slog of the Lambert Glacier Traverses in the 1990s, have been superseded by advances in remote sensing. The mass-balance of the polar ice sheets is now measured using satellite imaging, ice-penetrating radar for investigating deep ice and other measurements of mass variations. Estimates differ between technologies and research teams, and issues of measurement still need resolving. But significant losses are being recorded. Richard Kerr summarising in *Science*, 24 March 2006: 'some of the glaciers draining the great ice sheets of Antarctica and Greenland have sped up dramatically, driving up sea level and catching scientists unawares.'

Ice sheets are huge and complex. Mass variability is difficult to measure. Scientists debate the mechanisms driving ice flow. Eric Wolff, BAS glaciologist, in February 2007: 'we don't know polar ice's status. We don't know the processes that act to maintain, or destroy, it.' The models didn't predict the dramatic acceleration of glaciers.

Scientists worry about what they have missed. They can only guess at the implications.

In Greenland the accelerating glaciers appear to be affected by both surface melting and shelf-bottom melting. In Antarctica by far the smallest loss comes from the East Antarctic Ice Sheet spread over the high continental plateau. Scientists say they have not – so far – been able to determine whether the ice sheet is in balance or not. Most of the leaking is coming from that remote, geologically complex, difficult to access area where the limb of the Antarctic Peninsula beds into the continent. The groin, to continue the body analogy. Much of the West Antarctic Ice Sheet here is grounded on bedrock and sediment that is deep below sea level. Sectors are thinning, rapidly. The West Antarctic Ice Sheet enters the ocean in vast floating ice shelves. Scientists now understand that possible changes in the coastal regions of ice sheets must be included in calculations when monitoring their mass.

The coastal and near-coastal regions of the West Antarctic Ice Sheet are probably the most active in the Antarctic continent, with rapid increases in their discharge. The greatest uncertainty – especially in relation to predicting global sea-level rise – is most probably what is happening, now, here in the West Antarctic Ice Sheet; the potential for changes already underway in this massive wilderness of ice to accelerate. The most vulnerable parts, the area considered most susceptible to warming, could, scientists calculate, deliver enough ice off the Antarctic continent out into the ocean to raise global mean sea level by 1.5 metres: sections of an ice sheet experiencing sudden collapse. Antarctica starting to deglaciate. The wild card.

Recent events on the Antarctic Peninsula show that substantial changes can happen, fast. Time-scales can be much, much shorter than scientists thought. But trying to work out what might happen to the polar ice sheets in the current warming world is exceptionally difficult. Trying to work out what might happen in a future warming world is even more difficult. Should the destabilising of parts of the polar ice sheets be factored into future scenarios of a warming planet? Dismiss the possibility? Wait for further research? Are the time scales millenial?

Centuries? Less? If warming reaches the point where parts of the Antarctic continent should start to go – that is global. Antarctica is the giant that can shake the world.

James Hansen, Director of the NASA Goddard Institute for Space Studies: 'as with the extinction of species, the disintegration of ice sheets is irreversible for practical purposes.'

Moves from ice ages to not-ice ages, from what has been called hothouse to icehouse, and from icehouse to hothouse, have occurred in Earth's long, long history. The transitions happen over huge time scales. But what worries some scientists now is the way a warming world might be affecting the ways the large polar ice sheets – those highly significant remnants of our ice age world – maintain their stability. Their sensitivities to increasing rises in temperature are not known, or understood. Ice sheets are not uniform chunks of ice that melt, or don't melt. They are vast varied structures, with interiors and coasts, the size of countries. Beneath their mighty weight and height lie the contours of hidden landscapes – mountain ranges, valleys, old flood plains, deep dark lakes. All affecting the rate and rhythm of their progress. Yet still they move, inexorably, down from the high land, out to the sea, carrying everything that has arbitrarily landed on their bulk, abandoned buildings, meteorites fallen from space, dead explorers, fuel drums, route markers.

But time-scales can change. A flow of ice might slow, or stall. Or it might suddenly accelerate. Evidence can be extracted from small-scale glaciers undergoing rapid change in marginal areas of the world. But what of the mighty ice sheets? The ways their sensitivities work are not known.

The UN Intergovernmental Panel on Climate Change released its Fourth Assessment Report in three summaries for policymakers, from February to May 2007. Years in the making, drawing on the work of many scientists, the key findings have required the unanimous agreement of more than a hundred participating governments. The conclusions will provide the framework for international discussions for years to come. The Report lays out a bleakly realistic picture of climate change. Warming of the climate system is unequivocal. The world has little time to reverse the trend of rising greenhouse gas emissions.

The summary for policymakers from Working Group 1 gives a projected globally averaged sea-level rise at the end of the twenty-first century of somewhere between 0.18 to 0.43 metres. Little different from the Third Report in 2001. According to the Working Group 1 assessment, current global model studies project that the Antarctic ice sheet will remain too cold for widespread surface melting. It is expected to gain in mass due to increased snowfall.

The vast solemn bulk of the Antarctic continent, isolated by ocean from everywhere else, weighed down with its ice sheets as thick as Manhattan tipped sideways, is massively cold. Surface temperature records are sparse. There are too few to be statistically helpful: warming on the Antarctic continent has been considered to be no different from the global average. Lack of data, contributing IPCC scientists say, has excluded Antarctica from aspects of their forecasting. The usual elongated impression of the Antarctic continent appears across the bottom of some of the report's maps: everything below coloured grey, indicating 'no data'.

IPCC assessments have an inbuilt constraint. Scientists can only assess research already in the public domain; lengthy processes preclude the latest publications from the panel's conclusions. The scientific papers published in key journals reporting losses from polar ice sheets appeared too late to be included in the latest IPCC models, although the absence of the full effects of changes in ice sheet flow is acknowledged. Susan Solomon, leading US atmospheric scientist, co-chair of Working Group 1, in an interview published online in April 2007: the bottom line is uncertainty about the process in Antarctica. 'We don't know how to quantify it.' Nevertheless, 'we don't expect Antarctica to contribute much to sea-level rise in the next few centuries unless the ice dynamics effect were to become somehow much faster.'

But to me, travelling through a forest which is known to have tigers, is dangerous even if the number of tigers has not been measured, and their locations are unmarked. Lack of measurement does not remove the tigers. They still exist. When I suggested this analogy to one of the Working Group 1 authors, he replied – 'I have a headache.'

Some scientists, taking into account the recent losses from polar ice sheets, have publicly pushed the 2007 IPCC estimate for sea-level rise this century higher. Chris Rapley, director of BAS in April 2007: 'I would stress the uncertainties but emphasise that the IPPC estimate of up to 0.5 metres is almost certainly an underestimate.' Glaciologist Julian Dowdeswell, Director of Scott Polar Research Institute, a researcher at both poles: ice is dynamic at time-scales that are relevant to us, now. Changes are rapid. Sea-level is almost certain to rise at least half a metre this century. To Chris Rapley, the rise might be as much as a metre, and he quotes Stefan Ramsdorf from the Alfred Wegener Institute at a conference in Vienna in April 2007: it's a likely rise of between 0.55–1.5 metres.

Antarctica is a sensitive, critically important, pivotal part of the tightly coupled system that is our planet. Seeing the earth as an integrated system is new. Trying to understand Antarctica's role is very new. Pinning down any understanding of how the components of the climate system behave, and interact, in this geographically remote, physically isolated part of the world, is a struggle. The Antarctic continent is by definition a complex of responses, a mosaic of unstudied potential. Research is on-going. John King, in conversation in December 2006: what we are seeing right now is the climate system responding to a change in the composition of the atmosphere, occurring over a far shorter time scale than can have occurred naturally. But we don't know much about what is actually occurring on the continent, or about how changes in Antarctica might feed into the global climate system.

Extrapolating from the observed to the unknown is full of pitfalls. Confusion is easy. Bill warned me when I first talked to him in 1999, that people will want to apply the warming on the western side of the Antarctic Peninsula to all of Antarctica. He's right. The 3°C temperature rise reached only on the fastest-warming side of the fastest-warming part of a vast continent is regularly attached to the whole continent. As is the fate of Palmer's Adélies, transposed on occasion to all of Antarctica's penguins. The collapse of Larsen B ice shelf, that precisely located, specifically triggered event, does not represent the totality of Antarctica's ice. When I return from Antarctica people ask anxiously, but resignedly, 'has all the ice melted?' Where to start explaining? No. Antarctica's ice has not melted. If it had we'd be standing with dirty seawater sloshing around our ankles, and rising. Climate change has local, and regional impacts. The essence of the science I've described in this story is in the specifics. Data collected over sufficient time-scales, careful conclusions.

Earth's ice ages took the oceans and redeployed some of their contents as ice, releasing a greater surface area of land. Warming has reclaimed much of the ice as water, shifting it back into the oceans, making do with less land. But now warming is worming at the remaining ice. Knocking back glaciers, lessening the annual coverage of sea ice, shrinking the depth of frozen ground. Insinuating at the edges of the great ice sheets.

The specifics of what is happening in the polar regions have global implications.

Public perceptions of climate change have tumbled and eddied and swept forward, like branches carried by a river in a flood. Some have got stuck on snags or accreted detritus, but enough have travelled swiftly with the rushing water for an increasing acceptance. Our planet is irrefutably warming. No doubts, no buts. What has to matter is the climate, now, at this precise moment, with the living load our planet is currently carrying. And the speed of change. How fast.

Richard Alley, US polar geoscientist, speaking at the International Glaciological Society Symposium in Cambridge, August 2006: If you push too hard at the climate, something flips. People want to know. What does the future hold? When do we get in trouble? Everyone wants answers. They want predictions. We can't predict. But we can look at what will help us predict. Understanding ice, and measuring the mass balance of the ice sheets. With concentrated research effort we can do these things. That will tell us what will happen, and why. The ice matters.

For a long time I've been listening to ecologists, climatologists, meteorologists, geologists, oceanographers, palaeobiologists, palaeontologists; now, at this Cambridge symposium, I'm surrounded by people who are driven by ice – who study ice shelves, sea ice, glaciers, ice sheets, ice cores, frozen ground. The ones who work on the Antarctic Peninsula accept my description of the ferocious summer for the infamous season of 2001–02. The ferocious summer, they say, was like a river violently flooding. It permanently changed the river bed. It changed the surrounding country. A one-off – but it could happen again on the peninsula. It nearly did in 2004–05. A blocking high got established, then it broke up.

What has happened is that the frequencies have shifted.

To Richard Alley sea-levels have risen in the past. People dealt with them. We as humans can respond, effectively. And he pulled up a powerful image from deep in our cultures. God according to the Bible sent a rainbow to promise man that he would never again allow Earth to be flooded.

But I think of Palmer, in the ferocious summer. Rising temperatures sent a rare rainbow. A potent symbol, but potent in a different way. In high latitudes water comes from the sky packaged as frozen crystals, and stays frozen, as ice and snow. With increasing warmth, water gets delivered in liquid form, destabilising ice and snow and living things. As was happening at Palmer in 2001–02, that ferocious summer of rapid climate change.

Perhaps the Biblical rainbow isn't a promise, after all. It is a reminder.

Notes and Brief Bibliography

The information in this book comes mostly from interviews, conversations and observations. Some is part of the vast literature on climate change, or draws on the more specialised literature of Antarctica, including the writings of explorers and scientists. I have also relied on articles published in scientific journals and presentations and discussions at conferences and workshops, in particular the Antarctic Peninsula Climate Variability International Workshop held in Cambridge in September 2004 and Boulder, Colorado in May 2006, the International Glaciological Society Symposium in Cambridge in August 2006, the launch of the International Polar Year at the Royal Society, London in February 2007 and the Discussion Meeting on the science of climate change, 1–2 March 2007, also at the Royal Society.

Climate change in the Antarctic Peninsula, and Antarctica, has generated a large number of papers published in the scientific journals. Listing these sources would not be appropriate; but the papers in which Bill Fraser stated his hypotheses are, in order of first publication: W. R. Fraser, W. Z. Trivelpiece, D. G. Ainley and S. G. Trivelpiece, 'Increases in Antarctic penguin populations: reduced competition with whales or a loss of sea ice due to environmental warming?' (*Polar Biol*, 11, pp. 525–31, 1992); W. R. Fraser and W. Z. Trivelpiece, 'Factors controlling the distribution of seabirds: winter–summer heterogeneity in the distribution of Adélie penguin populations' (*Foundations for Ecological Research West of the Antarctic Peninsula*, vol 70 ed. R. M. Ross, E. E. Hoffmann and L. B. Quetin, pp. 257–72, 1996); W. R. Fraser and W. Z. Trivelpiece, 'The breeding biology and distribution of Adélie penguins: adaptations to environmental variability' (*Foundations for Ecological Research West of the Antarctic Peninsula*, pp. 273–85); W. R. Fraser and D. L. Patterson 'Human disturbance and long-term changes in Adélie penguin populations: a natural experiment at Palmer Station, Antarctic Peninsula' (*Antarctic Communities: Species, Structure and Survival, Scientific Committee for Antarctic Research, Sixth Biological Symposium*, eds. B. Battaglia, J. Valencia and D. W. H. Walton, pp. 445–52, 1997); W. R. Fraser and E. E. Hoffmann 'A predator's perspective on causal links between climate change, physical forcing and ecosystem response' (*Mar. Ecol. Prog. Ser.*, 265, pp. 1–15, 2003).

Aspects of recent Antarctic Peninsula research can be found in A. D. Rogers, E. Murphy, A. Clarke and N. Johnston (eds), 'Antarctic ecology: from genes to ecosystems. Part 1' (*Phil. Trans R. Soc.*, 362, 29 January 2007).

Significant research results relating to warming in Antarctica regularly appear in *Science*, *New Scientist* and *Nature*. Summaries of research, and interviews, can be found on the International Polar Foundation website www.sciencepoles.org.

Books on climate change include the very accessible Elizabeth Kolbert, *Field Notes from a Catastrophe* (London 2006) and Mark Bowen, *Thin Ice* (New York 2005). For clear overviews: Tim Flannery, *The Weather Makers* (London 2005); Al Gore, *An Inconvenient Truth* (New York 2006); and Robert Hensen, *The Rough Guide to Climate Change* (London 2006). For the science of rapid climate change: the Committee on Abrupt Climate Change, *Abrupt Climate Change* (Washington 2002).

For Adélie penguins and the ecology of the Antarctic Peninsula, the source book is David Ainley, *The Adélie Penguin* (New York 2002). Bernard Stonehouse (ed), *The Biology of Penguins* (London 1975) is still a classic. Parmelee's research life in the Palmer Archipelago is recounted in David Parmelee, *Antarctic Birds: ecological and behavioural approaches* (Minneapolis 1992).

Books on Antarctica: Antarctica is a wonderfully eclectic read with the journals of explorers, the musings of scientists, political analysts, books by ornithologists, historians and adventurers. Jeff Rubin's *Antarctica* (London 2000) is a comprehensive introduction to the continent, its history and science. But there is still nothing quite like the *Reader's Digest* publication *Antarctica: great stories from the frozen continent* (Sydney 1988). For more history, there's Alan Gurney's *Below the Convergence: voyages towards Antarctica 1699–1839*, (London 1997) and *The Race to the White Continent: voyages to the Antarctic* (London 2000). For scientists and naturalists writing about Antarctica the following have a devoted following: Louis Halle, *The Sea and the Ice: a Naturalist in Antarctica* (London 1973); Bill Green, *Water, Ice and Stone: science and memory on the Antarctic Lakes* (New York 1995); Peter Matthiessen, *End of the Earth: voyaging to Antarctica* (Washington 2003); Stephen Pyne, *The Ice: a journey to Antarctica* (Iowa 1986).

List of Illustrations

15. Larsen B Ice Shelf collapsing, 17 March 2002. National Snow and Ice Data Center and NASA, courtesy of Ted Scambos.

16. Adélie colony on Cormorant Island, 21 January 2003 (*Cara Sucher*).

17. Leopard seal, 2005 (*Cara Sucher*).

18. Adélie chicks close to fledging, Torgersen Island, 14 February 2002 (*Cara Sucher*).

19. Giant petrel and chick, Humble Island, 16 January 2003 (*Donna Patterson-Fraser*).

20. Norsel Point still attached by ice to Anvers Island, 15 January 2002. Photograph taken by flight observer on a Lynx helicopter from HMS Endurance, courtesy of Cara Sucher.

21. A new island, Norsel Point, January 2007. Photograph taken by flight observer on a Lynx helicopter from HMS Endurance, courtesy of Joe Pettit.

22. Author on Limitrophe Island, 1 March 2002 (*Donna Patterson-Fraser*).

23. The inner passages, 4 December 2005 (*Cara Sucher*).

Front cover image: Adélie penguins on Torgersen Island, 8 November 2001, their nest building material under snow. Ice covers the sea. Courtesy of Jennifer Tabor.

Back cover image: Adélies with chicks on Torgersen Island in January (*Heidi Geisz*).

Maps

1. Northern Antarctic Peninsula and Anvers Island inset – prepared by the British Antarctic Survey (p. xi).

2. Palmer Station and Islands – prepared by Environmental Research and Assessment Ltd, Cambridge. Data used by kind permission of NSF, SCAR and BAS. Main map data credits: rock coastline: Raytheon Polar Services Company surveys and USGS orthophotos. Ice coastline: British Antarctic Survey, Anvers Island and Brabant Island SQ19-20 3&4 Edn 1, 2005, BAS 250P Series, 1:250 000

Acknowledgements

From the moment I arrived in Antarctica, I was committed. I didn't expect it. Perhaps the nearest equivalent is the emotion some nineteenth-century Europeans felt for India, or Africa: a kind of continental commitment to every part of a complex, demanding, physical presence. I've been given generous chances to get there by organisations and individuals: the Australian Antarctic Division, Guy Guthridge and the Artists & Writers Program of the United States National Science Foundation Office of Polar Programs, Commander Tim Barton of HMS Endurance. Through them I've spent months in Antarctic waters on ice breakers and research ships, and months more ashore, in science stations and huts. This book is above all an acknowledgment of my gratitude for my opportunities to be in Antarctica, to absorb, learn and write.

When I began researching this book the debate was about whether climate change existed. Now it's about the nature of climate change – how warm, where, when, what is happening; and what can be done. Scientists were feeling their way. Now they are rock solid. The rest of us – the general public – are mostly still wondering how we got here.

The Ferocious Summer tracks one piece of climate change science, in one fast-warming place, and its impact on local inhabitants. I owe real gratitude to the people who have shared their experiences and thinking, shown and explained. At Palmer of course that means Bill Fraser of the Polar Oceans Research Group, without whom this book could not have happened, and I would like to thank him, very sincerely, and also Donna Patterson-Fraser and the rest of the seabird ecology team Chris Denker, Heidi Geisz and Brett Pickering, also Matt Irinaga and Pete

Duley from the 1998–99 season. Other scientists working at Palmer, or calling in, talked about their work with me, as did members of the support staff. I've generally avoided names in the text; people come to Antarctica to live their own lives. Station is their home. But managers Bob Farrell, and later Joe Petrit were unfailingly helpful; and the people who have shared their excellent photographs, in particular Cara Sucher, have been very generous. I thank everyone who contributed to my being able to tell the story of the ferocious summer at Palmer.

Polar scientists are particularly busy, often being away for months. I am very grateful to those who have found time to give interviews and answers, provide documents and signal articles, and access to conferences and workshops. I can mention only a few, but thank all: in the UK Chris Rapley, John King, Andrew Clarke, Tom Lachlan-Cope, Brian Gardiner, Dick Laws, Eric Wolff, Julian Dowdeswell; Richard Alley, George Denton, Warren Zapol, Ted Scambos, Christopher Shuman, Polly Penhale in the States; also Rear Admiral Ian Moncrieff, Stephen Cox, David Walton, Rob Massom and Peter Hooper. In Cambridge, thank you to the librarians at Scott Polar Research Institute, and the archivists at British Antarctic Survey; the mapping expertise of Adrian Fox, Peter Fretwell and Colin Harris; and the welcoming freedom to write, of my rooms in 'Penguin Palace' – once the home of the eminent British Antarctic explorer and administrator Sir Vivian Fuchs – at Wolfson College.

Andrew Clarke, John King, Bernard Stonehouse and Warren Zapol read all or part of the manuscript, giving it their necessary scientific attention, and I am very grateful indeed to each of them. Jean de Pomercu provided insights on an early version.

Working with my editor Peter Carson has been a real pleasure. I would like to thank Nicola Taplin, my copyeditor Matthew Taylor, and everyone at Profile Books for their help in getting this book smoothly to publication.

Finally I would like to thank Richard, my husband, for his good humour and constant companionship through a longer journey than either of us thought I was embarking on, when I first went to Antarctica.

Index

A Note on the Author

MEREDITH HOOPER's writing ranges from award-winning non-fiction books for all ages to academic articles, and her highly acclaimed fiction and information titles for children have been published in many languages. During the last fifteen years, selected as a writer on United States and Australian Antarctic programmes, she has specialised in writing about the history, geology and wildlife of Antarctica. She is a trustee of the Brussels-based International Polar Foundation, and the UK Antarctic Heritage Trust, and was awarded the Antarctica Service Medal by the US National Science Foundation in 2000. An Australian who arrived in the UK on a scholarship to continue her post-graduate studies in history at Oxford, she stayed, and now lives in London, though she returns regularly to Australia.